, *Vernal grass*; – 4., Avena flavescens, *Gold Oats*; – 5., Avena pubescens, *Downy Alpine Oat-grass*;
3r. arvensis, *Field Brome*; – 10., Cynodon, *Dog's Tooth Grass*; – 11., Cynosurus, *Dogstail Grass*;
5., Fest. duriuscula, *Fine-leaved Sheep's Fescue Grass*; – 16., Fest. pratensis, *Meadow Fescue Grass*;
Soft Grass ; – 20., Hordeum pratense, *Barley Grass*; – 21., Lolium perenne, *Perennial Ryegrass*;
Grass; 25., Poa trivialis, *Smooth Meadow-grass* ; – 26., Poa fertilis, *Fertile Grass*

THE
HISTORY
OF THE
WORLD
IN 100
PLANTS

SIMON BARNES

The Major Oak, *1844, by Henry Dawson (1811-78).*

THE
HISTORY
OF THE
WORLD
IN 100
PLANTS

SIMON BARNES

**SIMON &
SCHUSTER**

London · New York · Sydney · Toronto · New Delhi

MMXXII

CONTENTS

FOREWORD

We are the humans. We are the world-changers. We hold the planet in the palms of our hands. But we still eat the sun, as we always have done, as we always will for as long as we continue to exist. We can't eat anything else.

We consume the energy of the sun in the form of food, and the sun is made available for consumption by plants. Plants make food from the sun by the process of photosynthesis: nothing else in the world can do this. We eat plants, or we do so at second hand, by eating the eaters of plants, or, in longer food chains, the eaters of the eaters of plants.

Plants also give us the air we breathe. As they photosynthesise they take in carbon dioxide and push out oxygen. They direct the rain that falls and moderate the climate. Plants give us shelter, beauty, comfort, meaning, buildings, boats, containers, musical instruments, medicines and religious symbols. We use flowers for love, we use flowers for death.

We use the fossils of plants to power our industries and our transport. Across history we have used plants to store knowledge, to kill, to drive illegal trades, to fuel wars, to change our state of consciousness, to indicate our status. The first gun was a plant; we got fire from plants; we have enslaved people for the sake of plants. We have changed the world by planting some species of plant and by destroying others. We have used plants to understand how life on earth operates.

In most folk taxonomies and all supermarkets, fungi are considered much the same sort of thing as plants, even though scientists place them in a quite separate kingdom. In truth, fungi are more like us animals than plants, since they, too, are consumers of plants and wouldn't exist without plants. But in recognition of these older traditions I have put together nine chapters on fungi: these include *Penicillium*, which made modern life possible, and yeast, which has been used across the millennia to make life easier to bear.

We humans like to see ourselves as a species that has risen above nature, noble in reason, infinite in faculties, in action like angels. We have become as gods, doing what we will with the world.

But we still couldn't live for a day without plants. Our past is all about plants; our present is all tied up with plants; and without plants, there is no future.

Here are a hundred reasons why.

Opposite *A study in biodiversity 500 years before the word was invented:* Large Piece of Turf, *1503, by Albrecht Dürer (1471-1528).*

ONE
STRANGLER FIG

To sit in the shade and look upon verdure is the most perfect refreshment.

Jane Austen, *Mansfield Park*

It begins with a tree. Perhaps every story does. Let us ascend, branch by branch, our own family tree, climbing up through our ancestors, great above great above great. If we climb high enough – which means descending through time deep enough, counting in millions of years instead of centuries – we reach ancestors who spent most of their time in trees. They were smart, they had hands with opposable thumbs, which were great for grasping branches, and they had excellent three-dimensional vision, all the better to judge distances between branches. They lived in Africa and were admirably adapted for life in the trees. So far, so stable. Then the climate changed. That too will be a repeating theme in these pages.

When the climate changes, everything else changes, as we are in the process of rediscovering today. The world became a good deal cooler. The forests died back and the land became open grassland dotted with islands of trees. If you wanted to get from tree to tree, you needed to travel on foot. Humans began to walk upright, at first in order to reach the next tree more efficiently. And they found, in the course of advancing generations, that their grasping hands were suitable for using and making tools, their vision was as useful on the ground as it was in the trees, and their keen intelligence made them adaptable and resourceful. They became hunters and gatherers and the savannahs of Africa were their home.

But trees were still central to their existence, and not just as a source of food. Try walking on the savannah, as I have done many times. The environment provides food and water to those who know what they're doing, but hunting and gathering in the heat of the day is a fool's game. And whatever else our ancestors were, they were not fools. Their daily round involved a lay-up of at least four hours: out of the killing sun that overheats, exhausts and dehydrates. Every adult's mental map was based around shade trees. Trees that throw a deep, wide shade, with dense layers of branches overhead, have a great value to those who walk. When I escort visitors on walks in Zambia, we start at six. After three hours, the decision to rest up in the shade is greeted with rapture: to rest, to eat and drink, to talk, to think things over. It is an experience of unexpected depth and meaning for all who take part. It brings us back to our roots.

Sweet cool: ostriches and elands beneath a spreading strangler fig at De Hoop Nature Reserve, South Africa.

Of all the shade trees on the savannah, the strangler fig bears the crown. To walk beneath its shade is like entering a cathedral on a hot day's sightseeing: instant physical and spiritual refreshment, and a consequent sense of gratitude, even reverence.

The shade is impossibly wide; a mature strangler fig throws out branches in a radius of 65 feet (20 metres) from the trunk. You could rest dozens under these branches: families and wider social groups could come together and share the shade without imposing on each other. Here you could while away the hours: dozing, eating, drinking, singing, socialising, flirting, in time talking and planning. Human civilisation began in the shade of tree, and a strangler fig for preference.

There are various species of strangler fig in the savannahs of Eastern and Southern Africa, but they all have the same lifestyle. They evolved as forest trees: and forest life is about the competition for light. The conventional approach is for a seed to start on the ground; once germinated, roots grow into the earth and a green shoot grows upwards towards the light, seeking to outcompete all those around. The figs invented a way to beat the system.

There are about 850 species in the genus *Ficus*, including the familiar *Ficus carica*, the one that we eat. Figs have been cultivated for about 10,000 years; there is a case for claiming that they are the oldest cultivated plant. Pliny the Elder, the great observer and recorder of Roman times, noted that 'figs are restorative and the best food that can be taken by those who are brought low by long sickness'. As everyone who has eaten a fig knows, when you eat figs you eat a lot of seeds as well. Birds – in Africa especially green pigeons – eat fruit and later deposit the seeds with a generous dollop of fertiliser. The bad news for most fruiting plants is that often the birds dump the seeds in the tree they are perching in. But that is exactly what the fig wants. From the high branches a new shoot had only a relatively short distance to go to reach the light. As it sprouts a shoot, it also grows roots capable of taking moisture from the air.

So far so good. Many tropical plants have adopted this strategy, with aerial roots that never reach the ground. Such plants take nothing from the tree apart from its load-bearing ability: in tropical rainforests you can see trees festooned with plants, known collectively as epiphytes; in the moist air it's easy for the aerial roots to gather the water the plant needs. But the fig has greater ambition than these. Its roots keep growing downwards, and when everything goes well for the plant, the roots reach the ground. Once there the fig can start to spread itself. Soon it no longer needs the host for support; that makes it a hemiepiphyte.

The host tree is now held fast within a web of roots. At this mid-term stage, it's like looking at a single tree with two different kinds of leaves, two different sorts of flower and two different kinds of fruit. At this stage, association with a fig can even help the host, adding welcome support in times of storm. But eventually, if things work out, the fig will overwhelm the host, which will often die as a result.

The trunk that gave the baby fig tree such tremendous support will rot away and the fig will be left standing free, supported by a lattice of roots. It will be hollow on the inside, and easy to climb. No point in cutting it down: waste of a good shade tree, waste of a good fruit tree, and besides, the timber from all those straggling roots is pretty useless.

In season, the fig trees fed as well as sheltered the bands of human ancestors who sought out their shade. The trunk is also home to insects, bats, lizards and other reptiles, rodents, amphibians and birds. Fig trees often fruit at times when other trees all around do not, becoming an essential source of food for many species. This makes figs a keystone species in their environment: that is to say, a species that helps to maintain the environment it lives in.

It's not surprising that many traditional tales are told about as well as around the fig tree. A fig is crucial to the origin myth of the Kikuyu of Kenya; it is also a place that holds the spirits of dead ancestors. An infusion made from its bark is said to be helpful in pregnancy and makes for an easy birth; fig trees can also be used to make a poultice for wounds. The fruit is almost notoriously effective as a laxative in African as well as many other cultures. There is a charming tradition among the Nyanja people in response to the tight bond between the fig tree and its host: an infusion of fig bark can be used to reinforce the bond between man and woman, as a drink or even as a shared bath. There is an African proverb, relating to the fig and the small parasitic wasps that exploit the fruit: even if the fig tree is the most beautiful, it may have worms in its fruit – a useful principle for us all.

A fig tree also played a part in another story of origins. The fig species found in Southeast Asia fascinated the explorer-naturalist Alfred Russel Wallace, who called them 'the most extraordinary tree in the forest'. His study of the fig's struggle for existence helped him to an understanding of the mechanism by which life operates. He communicated his thoughts to Charles Darwin: and what Wallace had come up with was exactly the theory Darwin had been working on for the past quarter-century. Darwin, who had been postponing publication for years, knowing that he was sitting on a keg of dynamite, was at last spurred into action. Their joint theory was presented – to complete indifference – to the Linnean Society in London in 1858, but a year later, Darwin published *On the Origin of Species* as a solo project. That's when all hell broke out and Wallace, who had offered the only gentlemanly response to the issue of priority in the history of science, was reduced to a footnote. But without Wallace and his fig tree, Darwin might have sat on his theory for ever.

I have on many occasions sat beneath the shade of a fig, sipping tea, talking, not talking, just looking out upon the savannah beyond, grateful for the rest, planning what to do next, or as often as not just sitting. To sit beneath a fig tree is to make a journey back to the dawning of our species. Humanity was born in the shade of the strangler fig.

TWO

WHEAT

Give us this day our daily bread.

The Lord's Prayer

The world has been conquered by grasses from the genus *Triticum*. Fly over the agricultural countryside of the developed world, or just drive through Nebraska, and it becomes clear that humans are managing the planet for wheat. The crushed seeds of these grasses have nourished humans for centuries, for longer than we have cultivated them. For most of the world, they are not just an important food; they are food itself. Bread of course, but also pasta, noodles, semolina, bulgur, couscous, muesli, pancakes, pizza, cakes and breakfast cereals.

Seeds have always been part of the human diet. Chewing on sunflower seeds in their hulls unites us with our ancestors – and makes it clear that a fair amount of energy is required for a comparatively small reward. Better to hull the seeds in a mass: a more economical use of time and energy.

We modern humans differ from our ancestors by the musculature of our heads. We no longer have enormous jaws with powerful muscles; we no longer have a sagittal crest, to which still more powerful muscles could be attached. All this power was for processing food before digestion. Our more recent ancestors discovered more effective ways of getting nutrition: processing food before putting it in their mouths. They hulled, softened and moistened seeds, they tenderised plants by heating them and they tenderised the muscles of mammals and birds, breaking down the connective tissue that holds them together, again using heat. Starch residue 30,000 years old has been found on rocks by ancient human settlements: starch that came from seeds: seeds that have been crushed to make them easier to eat. At some stage the resulting stuff was cooked. You could make porridge with water; with a little less water, you could make a dough and heat that. It was perhaps humanity's first prayer, and the first prayer to be answered: give us this day our daily bread.

The greatest change in the history of humanity – the change from which all other changes followed – came around 12,000 years ago. Humans invented agriculture, became farmers, settled in a fixed place and established fixed communities. They gained security and a greater life expectancy in exchange for a lifetime of toil. Agriculture was invented more or less simultaneously in several

Winter defeated: The Harvesters, 1565, by Pieter Breugel the Elder (c. 1525-30 to 1569).

places across the world, but so far as Europe and the cultures of West Asia are concerned, it all began in the Fertile Crescent.

This is the land associated with the Nile, the Tigris and the Euphrates; on the map it looks more like a boomerang than a crescent. It was here, in these helpful circumstances – good climate, fertile riverine soil and easy access to water – that people stopped looking for food and started producing it. They domesticated animals and they domesticated plants. The most important plants were grasses of the *Triticum* genus. (A genus – a group of relatives – is one step up from species and one step down from family.) That is to say, wheat.

At some stage people started to use the pounded seeds to make leavened bread. The date for this invention is open to speculation, but you can recreate it in your own kitchen, as I often do myself. If you leave a porridge of water and pounded wheat – flour – it will gather wild yeast spores from the atmosphere and start to ferment. You can expedite this process with yoghurt, which brings in lactobacilli. Let this mixture develop, mix it vigorously with more flour, and you have a dough. Leave this dough a good few hours to rise and then cook it – and you have a loaf of sourdough bread, as ancient a cooked food as there is on the planet. Keep some of your fermenting mixture back, keep feeding it with flour and you always have the basis of bread to hand.

Wheat provides us with carbohydrates in the form of starch, which we use for energy, and a moderate amount of protein in the form of gluten, which we use for building tissue. (Modern strains of wheat are richer in gluten than their predecessors, and this causes problems for some people.) Gluten binds the dough, and is activated by kneading: the yeast releases carbon dioxide, which is held in the dough by the sticky gluten. You can feel the change in texture as you knead: the dough becomes bouncy. Gluten-free bread needs a binding agent like egg white to create pockets for the carbon dioxide. Bread also provides fibre. Flours with whole wheat (which includes the germ and bran from the seeds) have a good deal of fibre. Humans can't digest the fibre, but it adds bulk to stools and makes the human processing of food more comfortable.

Bread is central to the human cultures that use it: that is to say, more or less everywhere apart from the rice-centred areas of Asia. It lies at the heart of Christian ritual. The service of the Eucharist take the form of a symbolic meal of bread and wine. The bread represents – and to a Catholic actually is – the body of Christ, to be consumed in thankfulness.

The Roman satirist Juvenal poured scorn on people and politicians who look no further than gratification of appetite and cheap entertainment: bread and circuses, a phrase as relevant now as it was then. Pliny the Elder said: 'There were no bakers in Rome before [174 BC] and the war with King Perseus. The citizens used to make their own bread and this was the special task of the women, as it is even now in most nations.' The establishment of professional bakers was a step forward, for bread-making is a time-consuming business.

Agriculture was never a soft option. It was (and is) bitterly hard work, in the preparation of the soil, in the weeding of the fields, in defending it from pest species, in the gathering of the crop and subsequently in the processing of the seeds into flour and the manufacture of bread. Agriculture has always been vulnerable to the uncertainties of weather. Domesticating plants is a double-edged business, for there is always a payback. By planting a field of wheat, the owner has all the seed he wants. But he has also created vulnerability: for there is little resilience in a monoculture. As agriculture developed, it has increasingly moved away from diversity. Diseases, fungi and insects will all take advantage, easily moving from plant to plant, without unviable or unpalatable species to interrupt their progress. Traditional agriculture always flirts with disaster: to be warded off by prayers, hard work and anything else you can come up with.

Soil can be improved by the addition of fertilisers, initially the dung of domestic animals. Productivity could be increased by rotating crops: alternating wheat with legumes. Legumes include beans and lentils, also clover and alfalfa that can be grown as fodder crops: they fix nitrogen in the soil, and that benefits the cereals that you plant the following year. Crop rotation was invented at least 6,000 years ago. It also pays to rest the land occasionally: the book of Leviticus in the Bible instructs the Israelites: 'When you come to the land which I will give you, then shall the land keep a Sabbath unto the lord.'

The pace of agriculture hotted up in the eighteenth century, with the invention of a series of mechanical devices that made agriculture less labour-intensive, and therefore capable of feeding more people. The seed drill was actually invented centuries earlier, both by the Ancient Babylonians and the Chinese, but it wasn't in widespread use in Europe until much later. Before the seed drill was invented, seed was broadcast: a word that now has another use altogether. The haphazard nature of this business is caught in the biblical parable of the sower; the seeds lost to the farmer from a bad landing are compared to souls lost to the Lord from a bad life. But a seed drill planted seeds a uniform distance apart at a uniform depth, creating the regimented fields that we know today. The inventor of the horse-drawn seed drill was the eighteenth-century genius Jethro Tull.

Wheat's conquest of the world continued with the increasing industrialisation of farming and baking, and also with the increasing Westernisation of the world. Bread infiltrated all cultures, even the rice-growers. (A favourite snack in India is the toast sandwich.) The process was made easier by the invention of the combine harvester: a machine that combines the three great tasks of reaping, threshing and winnowing: that is to say, cutting down the corn, removing the seeds and then removing the seed husks. The first combine harvester was in use in the United States in 1835 and was pulled by a horse.

These days diesel-powered combine harvesters work in formation across endless prairies of wheat, blowing the husked seeds into trailers that travel the fields alongside them behind their tractors. The crops are protected from insects and other invertebrates, from fungi and from other problems by chemical means. The soils are treated with synthetically produced nitrates. Competing plants – weeds – are killed by pre-emptive use of herbicides.

The advances in agriculture have allowed the world to feed a great many more people; many more people now survive and breed in their turn. The world's human population continues to rise, and more and more food is required. Genetically modified strains are considered by some to be the answer, while opponents cite the law of unintended consequences, which have created ecological problems across the world.

Bread remains central not just to life but to the way we think about life. Marie-Antoinette is famous for a single remark, uttered, according to legend, when she was told that the poor people had no bread: '*Qu'ils mangent de la brioche,*' traditionally translated as 'Let them eat cake.' In fact, brioche is not cake; it's a bread made with a dough enriched with egg and butter: a luxury rather than a necessity. In 1917 the Bolsheviks promised the Russian peasants 'peace, land and bread'. What more could anyone want?

Bread is the staff of life, a cliché that dates back to the seventeenth century. In my hippy youth we referred to money as bread. Money is also dough: Bobby Locke, the South African golfer who won four major tournaments in the 1940s and 1950, famously said: 'You drive for show but you putt for dough,' meaning that mastery of the less glamorous part of the game is what makes you a champion.

Wheat is everywhere. World trade in wheat is greater than all other crops combined. In 2021 the world produced 772 million tonnes of wheat: and that's increasing. Wheat is the leading source of vegetable protein in most diets (wheat is about 13 per cent protein). Unsurprisingly, more land is employed for wheat production than any other crop. The planet earth has been modified and managed more or less from top to bottom for the convenience of a species of grass that grows wild in West Asia.

Opposite *Eye of the beholder: Marie-Antoinette with a Rose, 1783, by Elisabeth Louise Vigee-Lebrun (1755-1842).*

THREE
ROSE

Send two dozen roses to Room 424 and put 'Emily, I love you'
on the back of the bill

Groucho Marx in *A Night in Casablanca*

Should you find yourself in a church with your spirits flagging, you can turn to the Bible and read the most beautiful erotic poems ever written: 'I am the rose of Sharon and the lily of the valleys...' The Song of Solomon, sometimes called the Song of Songs, is about 3,000 years old – and when seeking to express perfect beauty (the narrator is female), the poet turns to a rose. Poets and lovers have been turning to roses to express love and desire throughout the three millennia that followed, and no doubt for thousands of years before that.

What is the evolutionary explanation for beauty? What is its survival function? Nuptial gifts are offered in courtship throughout the animal kingdom – spiders silk-wrap food items, penguins offer pebbles, marsh harriers pass gifts of food in flight. Did one of our female ancestors adorn herself with a rose? Did a male ancestor offer her a rose, to express both his admiration for her beauty and his own desire?

My luve is like a red, red rose
That's newly sprung in June...

That's Robert Burns and one of the greatest out-and-out love songs of them all. What does he mean by his love? Is he referring to the woman he loves and comparing her beauty to that of a rose? Is he talking about his own passion, now in full flower? Does the redness of the rose express the physical nature of his desire? Like any poetry that's any good, these simple verses have a dozen or more meanings: but no matter which way you pluck them, it's clear that roses and love are inseparable.

Wild roses grow in Europe, Asia, North America and Northwest Africa. I came across a wild rose garden in a valley bottom in Armenia: a stream flowed through it, leaving pleasant uncluttered banks, and it was clear that a bottle of champagne and a beautiful girl would make it paradise:

While the Rose blows along the River Brink,
With old Khayyam the Ruby Vintage drink...

Lines from *The Rubaiyat of Omar Khayyam*, in the much-loved Victorian translation by Edward FitzGerald. There are about 300 species of rose, but there are untold thousands of cultivars; 30,000 has been suggested. Species in the genus *Rosa* hybridise enthusiastically, giving immense scope to the gardener in search of a beauty still more perfect. Most wild species are white or pink, though there are some yellow and red. Most have five petals, so the changes wrought by cultivation have been immense: a cultivated Juliet rose is no more like a hedgerow *Rosa canina* or dog rose than a Yorkshire terrier is like a wolf.

Rose plants are woody and bushy. They produce fruit, normally known as hips, which contain the seeds. (Note that fruit are not necessarily edible: a fruit is the ripened ovary of a flowering plant containing the seed or seeds.) The flowers of wild roses are pollinated by insects; the seeds are dispersed by birds, who eat the fruit and spread the seeds in their droppings. Many domesticated varieties don't produce hips; many are too tightly petalled to be pollinated. Like many domesticated plants, most roses are incapable of propagating without human assistance. This is mostly done by taking cuttings: what is technically termed vegetative reproduction. The technique is familiar to all gardeners: a piece taken from a mature plant will, in helpful circumstances, put out roots and establish itself as a new plant, genetically identical to the parent plant: in effect, a clone.

When did people start cultivating roses? There is evidence dating back 2,500 years, but the idea of keeping plants for their beauty rather than their nutritional value is likely to go back much further: ever since the human elite had leisure, wealth and people to command. Perhaps humans have grown plants for their beauty for as long as they have cultivated wheat: and certainly what works for wheat works for less functional plants – weeding and manuring makes them bigger and better.

Roses were also cultivated for their scent. This can take the form of oil or attar of roses, which is obtained by steam-distilling crushed petals, and of rosewater, which is made by simmering petals in water. These covered up bodily odours in times when hygiene was less easily attained. Rosewater is used today in cosmetics and medicine; modern foods made with rosewater include baklava, halva, gulab jamun and Turkish delight.

Roses are so well-known and so greatly loved that they have grown rich in symbolic meaning. The rose was the flower of Aphrodite, the Greek goddess of love; in one story, wounded by the thorns, she stains white roses red. She anoints the body of the fallen Trojan hero Hector in oil of roses. She is painted surrounded by roses by Dante Gabriel Rossetti in his *Venus Verticordia*.

Roses were associated with the excesses of the Roman Empire: emperors bathing in rosewater, banqueting halls covered in rose petals, peasants forced to grow roses for imperial delight instead of crops to feed themselves. It is said that Cleopatra seduced Antony in a bedchamber paved with rose petals.

Despite – perhaps even because of – this licentious history, the rose also became a symbol of purity, associated with the Virgin Mary. The rosary, the Catholic aid to prayer, especially prayers to the Virgin, commemorates this equation. Its beads are linked in five series of decades, recapitulating the number of petals on a rose. The rosary can be used for contemplations of the sacred mysteries of Maria (note the five letters): the five joys of Mary are the Annunciation, the Nativity, the Resurrection, the Ascension and the Assumption. Dürer painted *The Feast of the Rosary* in 1506, in which Mary is distributing roses to surrounding worshippers. This association of Mary and the rose continues in church architecture with stained-glass windows in the shape of a rose, most famously the great rose window of Chartres Cathedral. There are sacred verses to continue this association: a hymn of 1420 begins:

There is no rose of such virtue
As is the rose that bare Jesu.

Geometrical gardens in the Islamic traditions centre on roses; two of the most important Sufi works are *The Rose Garden*, by Sa'di, and *The Rose Garden of Secrets*, by Mahmud Shabistari. Roses are found with our most elevated thoughts and with our most elemental.

Naturally the rose has been co-opted for badges and emblems again and again. It is the flower of St George, patron saint of England; it is the state flower in five of the United States; in 1986 President Ronald Reagan made the rose the floral emblem of the United States. In England in the fifteenth century the civil war called the Wars of the Roses was fought off and on for more than thirty years: on one side the House of Lancaster, whose badge was a red rose, and on the other, the House of York, with a white rose. The conflict was resolved in 1485 at the Battle of Bosworth Field; the victorious Henry Tudor of Lancaster, now King Henry VII of England, married Elizabeth of York, uniting the two houses. This was commemorated by the Tudor Rose: a red rose with a white rose in the middle, both of course with five petals.

The red rose is associated with socialism, especially after the 1848 French Revolution. A rose can mean what you choose it to mean: but that meaning is always coloured by the rose's association with beauty and moral virtue. But like all symbols, roses remain ambiguous. In *The Songs of Experience* William Blake portrayed a rose in distress damaged by harmful forces. Perhaps it's the best of all rose poems.

O Rose thou art sick.
The invisible worm,
That flies in the night
In the howling storm:
Has found out thy bed
Of crimson joy:
And his dark secret love
Does thy life destroy.

The cultivated rose was championed by Empress Joséphine of France, wife of Napoleon. She wished to turn her garden at the Château de Malmaison into 'the most beautiful and curious garden in Europe', and roses were at the heart of her vision.

Opposite *Two imprisoned knights listen to Emilia singing in the rose garden: an illustration to Giovanni Boccaccio's* Le Teseida, *published in 1468.*

The rose is not only a flower that pleases humans on a deep level: it is also remarkably receptive to all, a kind of ingenious manipulation. The range of the forms of modern roses is dizzying, and the ideas of beauty are frequently bizarre. The family garden of my childhood had a variety of rose called Blue Moon, with flowers a rather sick-making mauve. Disliked by the entire family, it bloomed defiantly year after year and was still hard at it when my parents moved twenty-five years later.

Roses remain the world's favourite cut flower: picked as buds and then refrigerated, to be flown all over the world. The need for roses, for their beauty and as tokens of affection, routinely conquers our concern for the ecological cost. They have been painted with careful accuracy by Pierre-Joseph Redouté (who did the job for the Empress Joséphine) and Henri Fantin-Latour, and with impressionist energy by Monet, Cézanne and Renoir.

On St Valentine's Day, roses are given to beloveds everywhere. Our need for roses to state our love has never waned. Umberto Eco, author of *The Name of the Rose*, said: 'the rose is a symbolic figure so rich in meanings that by now it has hardly any meaning left.' Roses are inescapable: as inescapable as beauty, perhaps. Should we ever escape the cares of work and family, we are told 'to smell the roses'. And we do.

FOUR
PEA

$A + 2Aa + a$

gives the series for the progeny of plants hybrid in a pair of differing traits.

Gregor Mendel

We seldom reflect, as we look approvingly at the cheerful green of the peas rolling on the plate between the fish and the chips or adding richness to the bowl of *mutter paneer*, that it was peas that revealed the mechanism by which life operates. And it's rum to think that Louis XIV, the *roi soleil* himself, was entranced when presented with the impossible luxury of a fresh green pea.

Charles Darwin showed us the way life operates in *On the Origin of Species*, published in 1859. He explained why; someone else had to explain how. Darwin showed that the forces of natural selection ensure that the life forms most suited to the place and time they live in will do better than the rest: live longer and make more of their own kind – and if they bear offspring with the same advantages, these will also prosper, and perhaps in turn pass on still more favourable traits to their own offspring. The most suitable – the most fit for the purpose – do best: survival of the fittest, selected naturally by the forces of the environment. Darwin's book set the world on a roar because it implied, though never stated, that humans must be part of the same system: that we must share a common ancestor with monkeys.

It was the logical inference from decades of intense study of living things, but there was (and is) continuing hostility to Darwin's conclusions. Among the myriad objections was the question: 'OK – bright idea. But how does it actually work?' This question occupied those who supported as well as those who rejected the idea of natural selection. Darwin worked on a notion he called pangenesis, which was brilliant but wrong.

So how does it work? The answer lay in the pea. It lay in the mind of a monk who was studying and performing experiments in St Thomas's Abbey in Brno, a vowel-deprived town in what is now the Czech Republic. He was pretty sure of his answer before Darwin asked the question: his essential work on peas lasted between 1856 and 1863. His name was Gregor Mendel and Darwin never once heard of him. His work on peas and the factors that govern inheritance was published in 1866 – and hardly anybody noticed.

How life works: Gregor Mendel, monk and pioneer of genetics, studying peas.

Mendel was born in 1822. He was educated in science, and was a brilliant mathematician, but as the son of a poor farmer he lacked the money to devote himself to education. So he became a friar, a choice that allowed him to live without the 'perpetual anxiety about means of livelihood', as he said. It also gave him the opportunity to do science. His baptismal name was Johannes; he took the name Gregor when he joined the Augustinian order.

He set out to study inheritance. Initially he worked with mice, but his abbot objected to the idea of a monk devoting his time to mouse sex. Mendel switched to plants, assuming rightly that the principles that govern inheritance are the same for plants and animals. He worked with a garden of 5 acres (2 hectares), studied about 28,000 plants, almost all of them peas – and he cracked it. Everyone has always accepted inheritance; it's the founding principle of the aristocracy as well as Darwinism. Mendel told us how it works.

He worked on seven different traits in different varieties of pea plants: height, pod shape and colour, seed shape and colour, flower position and colour. He noted what took place with meticulous care and made a series of calculations of classic precision and beauty. (So much so that it has been suggested that his results, with so little experimental error, are too good to be true. All the same his conclusions have been re-proved many times.)

Inheritance is not a simple matter of blending. If you cross a tall plant with a short plant, you won't get a bunch of medium plants. You get some tall ones and some short ones: in fact, three talls to one short, if tallness is the dominant 'factor', as Mendel called it. Put one of these resulting shorts to another short and you will get some short ones – and some tall ones as well, if the factor for tallness is recessive. In other words, Mendel discovered genetic inheritance. What he called factors we now term genes (or alleles, meaning groups of genes). He proposed three laws of inheritance: the law of dominance, the law of segregation and the law of independent assortment. They stand to this day.

The world throws up astonishing coincidences that startle even the most rational. In the same two-month period of 1900, sixteen years after Mendel's death, three different researchers in three different countries rediscovered Mendel's work and replicated his experiments. Not only that – they acknowledged his priority. Their conclusions reinforced those of Darwin, and in the 1930s and 1940s the work of the two men was viewed together as 'the modern synthesis' – or, in a term demeaning to Mendel, Neo-Darwinism. Darwin was helped to his understanding by breeding pigeons; Mendel worked on the equally homely pea.

Peas were grown in Egypt in the Nile Delta 7,000 years ago, but they weren't consumed as sweet green spheres that give to the teeth. They were eaten as mature seeds, which need a fair amount of soaking and cooking to make palatable. But they are good food, high in protein and fibre. People for whom meat was a luxury could keep going with peas. What's more, peas keep well, even in a hot climate. Dried peas, kept dry in a sealed container safe from insects, will keep until long after the next year's crop of peas has been harvested. Peas were a significant plant in the Middle Ages; we might even owe the survival of our ancestors and therefore our own existence to peas.

Pease pudding hot! Pease pudding cold!
Pease pudding in the pot, nine days old.
Some like it hot, some like it cold,
Some like it in the pot, nine days old.

The old rhyme, its origins long lost, tells us about a food that was never a treat, but the best possible standby – and it also gave the title to *Some Like it Hot*, the great film of 1959, starring Marilyn Monroe. Peas filled bellies and provided fuel for tomorrow: they kept hunger and even famine at bay. They were and are extremely useful crops, as part of the rotation with wheat (see Chapter 2). The roots of pea plants fix nitrogen in the soil: once the pods are harvested the remaining plants die and return nitrogen to the soil, making it suitable for wheat-growing again.

But peas had a vogue as a luxury item, for they can be harvested before the seeds are ripe. These immature seeds are sweet, brightly coloured and easy to consume. The drawback is that they don't keep very long: in warm weather (more likely than not when peas are harvested) they last only a few days – so fresh young peas could only be food for the rich man's table. Louis XIV had a passion for them. He grew many delicate vegetables in the gardens at the palace of Versailles, in the 9-hectare plot known as the Potager du roi, the king's kitchen garden. His brilliant court gardener Jean-Baptiste de la Quintinie was able to raise fruit and vegetables early in the season by using glass (at vast expense). The king loved the place and often went out there with his gardener. It was a homely idea, but they didn't grow peasant food: artichokes, asparagus, beans in their pods (what we call French beans) – and peas. Peas were associated with prestige: there was competition among the nobility to serve the first peas of the year. People are said to have died from a surfeit of peas.

Fresh peas reached a wider public with the invention of canning: you put the food in a sealed container and then heat it, killing the microorganisms within. The process was invented in 1809 by Nicolas Appert of France; demand for canned food was accelerated by the First World War and the need to feed soldiers at the Front.

The advance of frozen food technology, along with home freezers, made young green peas still more widely available, and what's more, they taste better than the canned variety. The technique was invented by Clarence Birdseye of America. He noticed that when fishing off the coast of Labrador in winter, a fish taken from the water froze as soon as it was pulled from the sea – and it was palatable, even tasty, months later if kept frozen.

If you freeze things slowly, it doesn't work nearly so well. Ice crystals form and rupture the cell membranes, and when the food is thawed the water runs out and takes the taste of the food with it. Birdseye established the technique of rapid freezing in 1824. You can buy Birdseye peas in the supermarket today and live like a king.

FIVE

WILLOW

And twelve yellow willows shall fellow the shallows…

Robin Williamson, The Incredible String Band

We associate willows with a certain grace and a pronounced fondness for water. They are most recognisable in the form of the weeping willow, much planted by lakes in parks. Cricket bats are made from willow, and the more pompous commentators refer to the bat as 'the willow'. Willows were useful to our ancestors for catching fish and making baskets. But willows can also ease fevers, aches and pains. Humans learned to synthesise the substance found in the sap of willow trees and it became a commercial product – aspirin.

There are about 400 species of willow, all in the genus *Salix*, including sallow and osier. Willows are found across the northern hemisphere in cold and temperate places, mostly in wet soil, where they are deeply comfortable, unlike trees in many other genera. They grow into tall trees, often with long, thin, flexible twigs, from which they get their flowing and graceful nature. In colder places where the growing is necessarily slower, they live as creeping shrubs. They put out flowers before the leaves come, in the form of catkins; male and female flowers on separate plants (technically, that makes them dioecious).

Willows are often deliberately planted along water courses where their roots add strength to the banks. This helps to prevent flooding and to keep waterways navigable. The roots grow enthusiastically and can then become counter-productive, so far as humans are concerned, by clogging drains, sewers and septic tanks. They sprout eagerly from cuttings, and will reproduce from their own fallen branches. Willows were imported to Australia to strengthen waterways, where they have out-competed native eucalyptus trees (see Chapter 92) and are classified as 'a weed of national significance'.

Willows produce a bitter bark sap in noticeable quantities. This is rich in salicylic acid, which is a painkiller and a reducer of fevers. The question of how people discovered this startlingly useful property haunts all such stories. Clay tablets from Ancient Sumer explain how willow helps with pain and fever; papyrus (see Chapter 20) from Ancient Egypt tells the same tale. Hippocrates, the Greek physician who lived in the fifth century BC, also noted the efficacy of willow-based medicine. Willow was part of the routine pharmacopeia of Roman and European medieval life. It was also used by populations in pre-Columbian America.

Healing properties: Weeping Willow, *1919, by Claude Monet (1840-1926).*

Willow was effectively rediscovered for the modern world by an eighteenth-century English clergyman named Edward Stone. The story is that he was taking a walk while suffering from ague, which is a catch-all term for fever. If he was out walking while suffering, we can probably assume that it was some kind of feverish cold, unpleasant but not life-threatening. In the course of his walk – perhaps the fever had rendered him slightly daft – he nibbled a piece of willow bark. He knew that the bark of the cinchona tree, which contains quinine (see Chapter 7), was also very bitter and wondered if this had a similar medicinal value.

He linked this possibility to the doctrine of signatures: if a plant looks like a body part, it must be therapeutic to that body part. That's why we have plants with names like liverwort, lungwort, spleenwort and even toothwort. Stone reckoned that since willows like wet places, and wet places are associated with ague, then the tree must have been put there to cure the disease. The doctrine of signatures is of course long discredited: but Stone was right for the wrong reason. So he dried and powdered willow bark and gave it to people suffering from ague, and it made them better. He wrote up his findings and they were published by the Royal Society in London.

Henri Leroux of France isolated the active ingredient, salicin, in crystalline form; in 1874 Hermann Kolbe succeeded in synthesising it. The problem was that in large doses, the stuff causes vomiting and even coma. Felix Hoffmann then came up with a synthetically altered version which caused less digestive upset and successfully treated his father's rheumatism. This was acetylsalicylic acid. The dye and medicine firm Bayer marketed it under the name of aspirin.

Aspirin was the world's leading painkiller until paracetamol was first produced in 1956 and ibuprofen in 1962. By then aspirin was being used for a different purpose: it helps to thin the blood. By 1948 aspirin was being prescribed for heart-attack victims: it is used today to prevent heart attacks and strokes. The Nobel Prize for medicine of 1982 was awarded to the researchers who found out how it works.

The willow tree also provides the only wood considered suitable for the making of cricket bats. The wood is light and tough and it readily absorbs the shock of impact. It resists splitting, it is soft but durable, and contains pockets of air that create the perfect implement for smiting. The sprung cane handle helps with the shock absorption, but it is the nature of the wood in the blade that allows the batter to propel a cricket ball 100 yards with apparently minimal effort. English willow is used for all premium bats; bats made from Kashmiri willow are reckoned to be inferior, being more fibrous and dense.

Willows produce long, soft twigs that can easily be bent into different shapes. Since prehistoric times willow has been used for fish traps, which can be placed in water facing the stream: easily in but not so easily out. You can then collect your catch and carry it home in a basket made from willow. Wicker is a term for all plant materials that will submit to being woven, and the skill to use it dates back at least to Ancient Egypt, and most likely a great deal longer. Wherever willows grow, they were the plant of choice for making containers. Willow is also good for fences to keep domestic animals enclosed, and to make houses for humans.

People have long had a purely aesthetic fancy for the weeping willow, with its long fronds usually dipping down to water. It was named *Salix babylonica* by Linnaeus, who invented the science of taxonomy, publishing the first edition of his *Systema Naturae* in 1735. He invented the two-name (binomial) system, and the name he chose for the weeping willow was a reference to Psalm 137, which includes the famous lines: 'By the rivers of Babylon, there we sat down, yea, we wept, when we remembered Zion. We hanged our harps upon the willows in the midst thereof.'

Weeping lovers: legendary couple Leyli and Majnoun beneath a weeping willow, Iranian textile, early twentieth century.

The rivers of Babylon were of course the Tigris and the Euphrates; there are no willows there, for here willow is a mistranslation of a species of poplar.

The weeping willow is native to North China and because of its pleasing nature, it has been cultivated for millennia, making it to Europe, like so many other things, along the Silk Road while it was active between the second and eighteenth centuries. Alexander Pope, the eighteenth-century poet, is said to have brought the weeping willow to England, begging a still-living twig from a basket of figs that had been sent to Lady Suffolk from Turkey. He planted this at his home in Twickenham, on the River Thames west of London, and all other English weeping willows are descended from it. This is not, alas, entirely true, but Pope's weeping willow at Twickenham certainly grew into a very fine tree.

Willows seem such thoroughly benign and helpful trees that it is almost tempting to believe in the doctrine of signatures or other philosophies of divine providence. But willows have more sinister associations. In willow-growing parts of China the people use willow branches to sweep the graves of their ancestors on the great festival of Qingming; in Japan willows are associated with ghosts. There are some sinister willows in English folklore, too, which have been rebooted in more recent works. The hobbits Merry and Pippin were imprisoned within Old Man Willow in *The Lord of the Rings*, and in the Harry Potter books, Harry and his friends frequently had trouble with the Whomping Willow.

SIX

GRASS

The days of man are but as grass.

The Book of Common Prayer

All flesh is grass. Words found in Isaiah, much relished and quoted with respect to the transitory nature of life: 'All flesh is grass and all the goodliness thereof is as the flower of the field.' Grass frames us. Grass once defined our world, and still does wherever we can manage it. We have converted endless acres of the world to grass. Humans have always relished the meat of grass-eating mammals, especially cattle. We sought their meat as hunters, then as pastoralists, then as farmers, now as city-dwellers. Cattle eat grass: many humans are eaters of grass at one remove. We are all grass, and whenever and wherever we can, we surround our homes with grass. We play games on grass or watch others play games on grass. All of life is grass.

Grasses include cereals like wheat, rice and maize, which together fill 51 per cent of human energy needs. We will take these on in more detail in separate chapters, along with bamboo, which, a little bafflingly to non-specialists, is also a kind of grass. For the purposes of this chapter, grass is the green stuff of lawns and meadows: the stuff we normally think of when we hear the word grass.

Grasses form the family *Poaceae*, which contains 780 genera and more than 12,000 species. They are all categorised as species of flowering plant, which is again slightly baffling to non-specialists. But grass will flower, given half a chance: it's just that the flowers are subtle and discreet; they don't require insects for pollination so they don't need to make a show of themselves. They are wind-pollinated, and in open grasslands, there is usually wind of some kind as well as plenty more grasses in flower for the randomly broadcast pollen to find.

So far so good. But grass has a USP, and that has helped to make the world what it is today. Grass doesn't grow from the tip; it grows from the bottom. The growing bit – the place at which cells divide and growth can take place – is technically the meristem: and in a grass species you find it near the bottom. That may not seem the most inspiring fact you've ever heard, but it's a world-changer all the same. It means you can chop the head off a stem of grass as many times as you like and it'll keep growing.

This strategy defends the plants against grazing animals, because they can keep going after they've been munched. It also makes the life of a grazing animal

Caption to come

How's that: cricket – the meadow game – played on Hampton Court Green, 1836.

possible. You can see the principle in action at the great Serengeti migration: the wildebeest munch and move on. They follow the rains, but they give the rains a good head start. By the time they get there the rain has already inspired a rich new growth of grass. Any other plant that puts its head up will also get munched: the difference is that the grass survives. You could say the wildebeest were managing the land for their own advantage; you could say that the grasses were exploiting the wildebeest in order to dominate the ecosystem.

The first humans hunted and gathered across grasslands of Africa: savannah is mostly grassland with islands of trees. Africa still has the biggest biomass of large wild mammals in the world, and that's because the grasslands support big herbivores and these support the great carnivores.

Pastoralism was a great leap forward for humanity: a semi-nomadic life based around herds of cattle, sheep and goats. There are two theories of how it began. One is that it was a natural development from hunting: good hunters know their prey and follow the herd, and the beginning of domestication is a logical development of that intimacy. The second suggests that pastoralism developed from mixed farming, as a way of exploiting land less helpful for the growing of crops. Like many apparently conflicting theories, the chances are they are both right. Pastoralism has been followed for millennia and in many places still continues. It is at base a leisurely exploitation of grass. When the grass has been eaten off, you move on. The free-grazers of the American West lived in precisely that way.

The establishment of civilisation was largely a process of cutting down forest to make grassland. Once that was done people could graze the animals they used for meat – and also horses that were the world's transport system for five millennia. The process of swapping forest for grassland continues, in a manner that began to trouble the world in the 1980s; forests being a great deal more effective than grassland at storing carbon and so slowing down the rate of climate change.

In the UK and elsewhere in Europe, much of the grazing was done on shared land, common land, but then the land was enclosed, for the benefit of the rich and powerful. A rhyme from the seventeenth century sums up this change:

> The law locks up the man or woman
> Who steals the goose from off the common,
> But leaves the greater villain loose
> Who steals the common from the goose.

Enclosed grasslands are now intensively managed for the grazing of domestic animals. This requires the introduction of fertilisers, fungicides and selective herbicides. The management also requires selective sowing of certain types of grass, ryegrass in particular. This is the process that gives pastures their bright uniform colour; such a field, more or less a ryegrass monoculture, is usually referred to as improved grassland.

Grass is also grown to feed animals who aren't actually on the field. Crops of hay – dried grasses – are winter feed for many domestic animals. But hay is a notoriously fickle crop, since it must be dried in the field. It is vulnerable to shifts in the weather; many a hay crop has been lost to a sprightly summer thunderstorm. Such disasters can be pre-empted by modern processes of cutting grass and allowing it to ferment: this is silage and it feeds cattle and other herbivores in winter. It has long been common practice to keep cattle in barns in the winter; they are increasingly kept in barns full-time and all the food they ever see is in a trough. This can involve food other than meadow grasses: we will meet some of these later on.

But grass is more to us than nourishment for the animals we eat. It seems to be essential to the well-being of humanity. We surround ourselves with grass. The UK is full of houses that have gardens and most have a patch of lawn, mown every week in the growing season. Suburban America is full of houses set well back from the road with an unfenced stretch of grass leading down to the road. Generally speaking every plot is identical and every one must be mowed. Housing estates in many places will have grass verges. The sides of major roads are often grassed over, inspiring Roger McGough's two-line poem to a beautiful petrol pump assistant on the M1:

I wanted your soft verges
But you gave me the hard shoulder.

If you are a person of colossal wealth, you will own a country house with a deer park. You will look out from your Palladian windows on the pleasantest vista any human can imagine and money can buy: a broad stretch of closely cropped grassland studded with mature trees leading down to water. The deer may or may not be in sight, but their presence is implied by the short sward. Why do we choose such a view? It's a fair assumption that what the rich actually get is something that the rest of us really want: why do they choose such a vista?

Cast your mind back to the opening chapter. When our earliest human ancestors sat under a strangler fig in a rare time of peace and content, they would have seen much the same view: a plain that supports large mammals, with fine shade trees and a place where you can get a drink. I suspect that this desire for an expanse of grass is an atavistic need: an attempt to create an idealised version of the landscape where the first humans thrived.

Every expanse of grass must be controlled. If you don't keep deer and you don't keep mowing it, it will grow. It will look, perish the thought, 'untidy'. It will tell the world that you have failed in your job of keeping nature in its place. The cut lawn represents both our need for nature and our fear of its power. When you see an unkempt front lawn you might speculate that the people in the house behind it are alcoholics: people who have lost control of their lawn and their lives.

For some, mowing the lawn is a necessary chore. For others it is a passion: lawns must be treated with selective herbicides, any upstart plant that gets in there must be eradicated, and the lawn must show a pattern of stripes: neat and, of course, lifeless. All the same, a mown sward is the most marvellous place for play. Its openness gives range to a great number of possibilities, and when you fall over, it doesn't hurt. Childish rough-and-tumbles developed into more serious games for young people, and so into the organised games we have today. Football is played on grass, in the forms of soccer, gridiron or American football, rugby union and rugby league and Australian Rules. Cricket, tennis, golf and baseball were all invented for grass. In cricket the nature of the 22-yard strip of grass between the two wickets is intensely variable and its condition alters from one day to the next in a match that can last five days. Golf is another game about grass: the grass of fairways and putting greens is so intensively managed that it no longer looks like grass, while the division between fairway and rough is like the division between wilderness and civilisation.

Some games have moved away from their grass roots: tennis began as a lawn game but is now more often played on courts of concrete and clay. Hockey – field hockey – is now mostly played on artificial turf. But sport has seized the imagination of every country in the world, and it began on grass.

Humans are a species of grassland animal. We have moved a long way in the 4 million years since we first walked the savannahs, but we retain our profound identity with grass. The short sward is where humanity began, and it is to the short sward we return, to please our eyes, to play our games, to assert our control and to soothe our souls.

SEVEN

CINCHONA

*Ship me somewheres east of Suez, where the best is like the
worst...*

Rudyard Kipling, 'The Road to Mandalay'

The cinchona is the tree that made empires possible. Without the cinchona tree the European adventures overseas would have been mere trading posts and military garrisons rather than the imposition of one civilisation on another. The cinchona has probably saved more human lives than any other plant on earth. As part of the same process, it provided the go-to alcoholic drink for a large part of the world. The cinchona was for 300 years the world's only source of quinine, which both cures and prevents malaria.

The cinchona (mostly pronounced sin-koner) is native to South America, where malaria was most probably unknown until the Europeans started arriving from 1492 onwards. The pathogens that cause the disease were likely transported in the bodies of European travellers, or those of their African slaves. They entered the ecosystem when these people were bitten: the humans infected the mosquitoes.

The tree's usefulness against fevers was widely known in local cultures, but the discovery of its power against malaria first required the introduction of the disease. Linnaeus named the genus *Cinchona* in 1742; there are around twenty-three different species within it. He chose the name from a story that the Spanish viceroy in Lima, the Count of Chinchón, had been cured of malaria by the bark of the cinchona tree. (He changed the spelling to Latinise it, so it would fit more tidily into his Latin-based binomial system.) In another version it is the countess who was cured, after which she brought large quantities of the bark to Europe to treat the afflicted.

Malaria was familiar throughout Europe, often known as ague (see Chapter 5), tertian fever or marsh fever. Mosquitoes breed best in marshland and the disease was associated with the air of the swamps; *mala aria* means 'bad air'. It was soon clear that the cinchona bark was very valuable stuff indeed. Its use in Western medicine was largely pioneered by Jesuit physicians and it was known as Jesuit's bark.

The subsequent history of the bark is confused and contradictory, full of discredited stories; it's as if we were talking about the furtive development of a

Dodd delt.

Jesuits Bark.

Prattent sculp.t

magical process. However, the bark made it to Europe, perhaps thanks to the generous heart of the cured countess. It was taken up by, among others, Robert Tabor (or Talbor), a British physician practising in the Essex marshes, and he was successful in curing malaria locally. He made his name by curing a British naval officer, a feat that came to the attention of King Charles II. (In some versions he cures the king himself.) Tabor then went to France and achieved wealth and fame, curing the heir to the throne or, in some versions, Louis XIV himself.

As the pace of European exploitation of distant countries increased, so did the value of the bark. Many trees were destroyed so that the bark could be stripped and exported. The Spanish colonialists, most active in South America, had easy access to cinchona. What other colonial nations wanted was a supply of their own, but by the beginning of the nineteenth century, its export was forbidden.

Seeds were smuggled out of Bolivia by Charles Ledger, whose assistant, Manuel, was beaten to death in the process. The British foolishly turned this booty down, but the Dutch bought it and established plantations in Java, then a Dutch colony. Clements Markham later established British plantations in India, in the Nilgiri Hills around Ootacamund, and in Sri Lanka.

It is important to understand the extent of malaria. The mosquito was not recognised as the vector for the disease until the beginning of the twentieth century, so no one thought that avoiding bites was the most important first step in avoiding the disease. Malaria has been a significant fact of human life ever since humans stopped the nomadic hunter-gathering and became farmers 12,000 years ago.

It's been reckoned that half the humans who ever lived died from malaria. Even today a conservative estimate from the World Health Organization puts the annual deaths at half a million. And it is not just the deaths: the routine of illness – a person in a bout of malaria is too weak to lift head off pillow – costs innumerable working days. If you look at an empire as a purely commercial concern, malaria affected turnover and profits to a devastating degree. Africa could not be fully exploited because of the prevalence of malaria: the continent was known as the White Man's Grave.

The establishment of cinchona plantations changed all that. Malaria could be cured: and it could also be prevented by regular doses of quinine: that is the alkaloid – a naturally occurring organic compound – extracted from the bark. (The name comes from a word for the tree in a Peruvian language.) This was first extracted from the bark in 1820, by Pierre-Joseph Pelletier and Joseph Caventou. The availability of quinine opened the colonies up to wives, children, family: colonies became domestic rather than purely commercial and military. In India, the British rulers established families that went through generations and called

Opposite Miraculous tree: cinchona in a hand-coloured engraving, 1795.

India home. The process of deep cultural exchange between two cultures has continued ever since, both celebrated and resented.

The problem with taking your daily dose of quinine is that it's oppressively bitter. The British in India took to mixing quinine with soda water and sugar. That was much better, but there was still scope for improvement – so they added gin. The first commercially produced tonic water came on sale in 1858. For a century or more the British ruled India by means of gin and tonic. Modern tonic waters are much lighter in quinine; most gin-drinkers prefer their bitterness subtle. Quinine fluoresces under ultraviolet light: you can reproduce this effect in bright sunlight if you examine your drink against a dark background.

During the Second World War the Americans lost their access to quinine when the Japanese took over Java and its cinchona plantations. The United States sent expeditions into South America in search of an alternative source; meanwhile tens of thousands of United States servicemen died from malaria in Africa and the South Pacific. Quinine has always had a military as well as a commercial value.

But that finally changed in 1944, when quinine was produced synthetically. Other forms of treatment have been developed since then. The tree that was central to the hopes and ambitions of conquering the world could now go back to being a tree. For 300 years it was one of the most important plants in the world: it now remains, like colonialism, a part of history that is hard to understand completely in the twenty-first century.

All made possible by quinine: a British banquet at the palace of Rais in Myenere; engraving by Henri Théophile Hildebrand (1824-97), published in 1882.

EIGHT
SUNFLOWER

I find comfort in contemplating the sunflowers.

Vincent van Gogh

Sunflowers were cultivated in North America long before Columbus crossed the Atlantic. They were grown for food, but no doubt their dramatic appearance was part of the appeal. Certainly it was the look of the plant that prompted Europeans to take sunflowers back across the ocean. Once there sunflowers became a crop plant all over again, useful and humble. But now the plant is ineluctably associated with the cult of genius and the legend of the tormented artist.

There are seventy species in the genus *Helianthus*, but it's the cultivated species *Helianthus annus* that mostly concerns us here: the one with the flower-head that looks like the sun. It's not technically a flower but an inflorescence. Each head comprises many individual flowers; each of the outer flowers, which most of us refer to as petals and a botanist calls ray heads, are in fact individual flowers. These outer flowers don't do sex, being sterile: they are a come-hither signal to insect pollinators, which feed from the many tiny flowers arranged in cunning spirals on the central disc. Sunflowers are famously tall: a good average of 3 metres (10 feet), while the record is 9.17 metres (over 30 feet).

They were cultivated for their fruit, which contain the seeds. What we refer to informally as sunflower seeds, with the husks that we strip away with our teeth, are technically fruit; what we actually eat are the seeds that lie within the fruit. The plants have been cultivated for 5,000 years; the seeds could be used as a crushed meal for flatbread, or mixed in a porridge with beans, squash and maize.

Sunflowers were taken to Europe for their beauty and were especially popular in Russia in the eighteenth century. The Russians discovered that a useful oil could be obtained from the crushed seeds. This was better news than it sounds: most oils were forbidden during Lent by the Russian Orthodox Church, and the discovery of sunflower oil legitimised a number of small pleasures.

Sunflowers grow well in places with plenty of sun. The oil was used in cooking, and the waste product, the crushed seed, was fed to livestock. Sunflowers were always handy things: but their startling appearance gave them meaning beyond their function. There was a compelling idea that sunflowers track the sun: turning their faces to the morning sun and following it across the sky all day until it sinks down in the west. William Blake wrote in *Songs of Experience*:

Ah sun-flower! weary of time,
Who countest the steps of the Sun:
Seeking after that sweet golden clime
Where the travellers journey is done.

But it's only while the sunflower is in bud that it countest the steps of the sun: once it's a full flower (or inflorescence) it holds still. That doesn't (or does it?) detract from a great poem about longing for some new age or place where we can live in freedom.

Sunflowers became increasingly useful as crops. One species, *Helianthus tuberosus*, has been cultivated, as the name suggests, for its edible roots, the food we call Jerusalem artichokes. The oil has become useful in many forms of cooking, baking as well as frying, not least because it is cheaper to produce than olive oil (see Chapter 82). It can also be used in biofuels; there is continuing debate about whether biofuels are a useful exploitation of the earth's resources (more on this in Chapter 95). The fuel company BP uses a stylised sunflower as a logo, presumably to distance itself in our minds from the equation of fossil fuels (the company's main concern) and their part in the continuing disaster of climate change.

Sunflowers were important to the Aesthetic Movement, and its doctrine of art for art's sake, which is normally dated from 1870 to 1900. It was about turning away from the values of increasing industrialisation, in which all humans were seen as parts of the great machine. The values of the Aesthetic Movement affect thinking today: for example in the notion that every home should be beautiful as well as functional. The movement was much taken by the simplicities of Japanese art. Chrysanthemums are important in Japanese paintings, and sunflowers look fairly similar, so sunflowers became a favourite flower, and even a badge of the new doctrine, which was of course much mocked. An exquisite young man with a sunflower is portrayed on a song sheet with the title *Quite Too Utterly Utter*.

In 1887 in Paris there was a small exhibition of avant-garde paintings. This included studies of sunflowers lying on the ground, painted with the most extraordinary commitment. The artist was Vincent van Gogh. Another painter, Paul Gauguin, liked them a lot, and later agreed to swap one of his own paintings for two of the sunflower studies. He said they were 'completely Vincent': the first of countless millions to reach that conclusion.

Van Gogh had a dream of setting up an artists' colony: a brotherhood of art, beauty, friendship and mutual support. Gauguin agreed to join him and van Gogh rented a house in Arles in the south of France. He got it ready for Gauguin's arrival by painting more sunflowers for him, this time in a vase. In one of van Gogh's terrifying explosions of creativity, he produced four studies of sunflowers in six days, in a great binge of paint, coffee and booze.

But let us not go too close to the Vincent Legend, in which van Gogh was a mass of blind instinctive energy possessed by that queer thing genius. He was consciously following the Dutch tradition of flower painting, but doing so in a new way and with the latest technology. Van Gogh loved yellow (see Chapter 49 on foxgloves) and for him it represented joy: and it was with joy that he turned to thrilling yellow paints that were newly available to him. The most famous sunflower studies use almost no colour that isn't yellow.

He chose sunflowers because of their commonplace nature; the fact that they are grown in their thousands in enormous fields, rather than as rare hothouse blooms. Beauty and joy should be matters of everyday life, accessible to all: and that was the foundation of van Gogh's art. There was also a practical reason: sunflowers are available cheaply, if not for free – fallen by the wayside, or easily

Vincent's flowers: Four Withered Sunflowers, *1887, by Vincent van Gogh (1853-90).*

borrowed from a field. 'Normality is a paved road,' he once wrote. 'It is comfortable to walk on but no flowers grow on it.'

So he wrote from Arles to his brother Theo: 'I am painting with the gusto of a Marseillaise eating bouillabaisse, which won't surprise you when it's a question of painting large sunflowers.' Many of the sunflowers he painted were past their best, wilting, losing petals (ray heads if you prefer). He was intrigued by the way that each one faded in a different way.

Gauguin came to Arles and after nine crazy weeks the artists' colony was at an end – though not before Gauguin had painted an affectionate study of van Gogh at work, painting... well, sunflowers, what else? The terrible rows, the self-mutilation, van Gogh's subsequent alternations of joyous masterpieces and personal despair, the awful end: his story is well enough known to leave it there... always with the thought that if he'd only hung on for just a few more years, he'd have known about the adulation and love his work now inspires. Even in the year of his death, there was a furious row at an exhibition in Brussels. Henri de Toulouse-Lautrec and Paul Signac both stood up for van Gogh against Henri de Groux, a Belgian artist who didn't care to be associated with a 'laughable pot of sunflowers'. A fist fight was narrowly avoided; in some versions of the story Lautrec challenged de Groux to a duel.

But van Gogh's sunflowers are now the world's joy. An estimated 5 million people gaze on the various versions of his sunflowers every year: van Gogh painted five in Paris and seven in Arles, one of which was destroyed by the American bombing of Japan during the Second World War. In 1987, a Japanese insurance magnate, Yasuo Goto, paid £25,087,500 for one version.

The sunflower paintings from Arles are as recognisable as the *Mona Lisa*. Each one is so often seen as a Great Painting that the flowers themselves tend to get lost: reproduced a million times in a million different forms, T-shirts, tea towels, fridge magnets and all. But it's not the biography or the mythology of the tormented artist that matter. It's the paintings: paintings about the overwhelming nature of joy and the awful fragility of such heightened experience.

NINE

OAK

Every oak tree started off as a couple of nuts who stood their ground.

Henry David Thoreau

The English have a thing about oaks, and they think it makes them special. But the oak is the national tree of many countries: for example Bulgaria, Cyprus, Estonia, France, Germany, Moldova, Jordan, Latvia, Lithuania, Poland, Romania, Serbia and Wales. In 2004 the oak became the national tree of the United States. In the Second World War English soldiers fighting for the land of the oak tree were up against an enemy that rewarded its heroes with the Knight's Cross of the Iron Cross with Oak Leaves. Many nations identify themselves with oak trees.

There are around 600 species in the *Quercus* genus, which is actually part of the beech family. The word oak is also used informally for many trees outside that genus, only some of which are related. There are oak species with unmistakable acorns in tropical rainforests. Even in England there is more than one oak. The so-called English oak, *Quercus robor*, is the one dear to English hearts, but there are also sessile, downy, turkey, cork, and holm oaks, some of which have been introduced to England. A certain kind of English patriotism is based on notions of purity of lineage: those that find the oak an appropriate symbol of such things will be interested to learn that oak species readily hybridise.

In much of lowland Britain the oak is the climax vegetation. Plants operate a succession: open ground giving way to brief annual plants, with more robust perennials taking over, after which the place will scrub up. The brambles that sprout up often act as natural tree-guards for the pioneer trees like silver birch. Agriculture and gardening are simply methods of controlling that succession; left to its own devices, your garden will turn into a closed-canopy oak forest in a few centuries.

The extent to which this actually happened across Britain is a matter of debate among historical ecologists: before widespread human settlement, wild cattle (aurochs), horses and deer, along with the rootling wild swine, must have kept many places relatively open: perhaps our ancestral landscape was a mosaic of open areas and stands of oaks. But whichever way you look at it, the oak is a climax species. It is also a keystone species, in that it changes the nature of an environment, dominating the landscape and offering opportunities for many more species to make a living. No English tree supports more species than an oak.

Trees to live with: oaks dominate The Rainbow Landscape, *c. 1636, by Peter Paul Rubens (1577-1640).*

In many places there are traditions of sacred oaks. Oaks impress with their size, their strength, their robustness and their age. There is a story told of Staverton Park in Suffolk: when the monasteries were being dissolved in the sixteenth century, the monks of Staverton begged a last favour – could they plant just one more crop, and leave the monastery when it was gathered in? This was agreed to – so the monks planted acorns. Certainly Staverton has a rich gathering of oaks half a millennium in age.

Oaks were sacred to both Zeus and Thor, gods associated with thunder. Oaks, standing proud of the rest, are frequently struck by lightning. Pliny the Elder noted the association of Jupiter (the Roman Zeus) with oaks, and said: 'Trees were the temples of the gods following old established ritual; country places even now dedicate an outstandingly tall tree to a god.' Oak groves were also sacred to druids.

Oaks were a useful food resource for humans. Acorns could be dried and ground into flour, a useful fallback when cereals were scarce. Acorns have been used to make a coffee substitute in times of wartime shortages. But it's been more usual to feed acorns to pigs. Allowing pigs free-range foraging in autumn woods is an ancient tradition: in many places the rights of pannage have been protected and disputed for centuries. In some areas a system of forest pastures has developed, supporting grazing animals and, in season, pigs. In Spain and Portugal the *dehesa* landscape is built around cork oaks: the bark itself is harvested for stopping bottles. Like many ancient agricultural practices, it has incidental benefits to wildlife, but the recent practice of selling wine in screw-capped bottles puts these at risk.

Stories accumulate around great oaks. There is a tree called the Crouch Oak in Windsor Great Park in England: it is said that Henry VIII and Anne Boleyn, soon to be his second wife, danced around it; later his daughter, Queen Elizabeth I, picnicked beneath its already inviting branches. The Major Oak in Sherwood Forest is at least 800 years old, and, according to tradition, Robin Hood hid from the Sheriff of Nottingham in its branches or he feasted with his Merry Men beneath it – perhaps both.

The Royal Oak stands in the grounds of Boscobel House in England: here the future Charles II hid after the Battle of Worcester in 1651; a Parliamentarian soldier walked directly underneath it without spotting him. Charles II himself told the story to the great English diarist Samuel Pepys:

> He told me that it would be very dangerous either to stay in the house or go into the wood (there being a great wood hard by Boscobel) and he knew but one way how to pass all the next day and that was to get up into a great oak in a pretty plain place where we could see round about us for they would certainly search all the wood for people that had made their escape. ... [We] got up into a great oak that had been lopped some 3 or 4 years before and so was grown out very bushy and thick not to be seen through. And there we sat all the day.

But no matter how wonderful a living oak might be, dead oaks were still more useful to an ambitious civilisation. Their wood is hard and dense, and stuff made from it is durable. I lived for fourteen years in an oak-framed house: it had stood for 500 years and is still going strong. Timber-framed houses were the basis of European life for centuries. Oak makes furniture with little delicacy, but it lasts. Oak is still used for barrels for the best drinks; here they can develop because they have contact with the air: to this day wine, brandy and whisky are matured in oak barrels.

But the use of oak in shipbuilding has done most to create the tradition of self-identification with oak trees. Viking longships were built from oak more than 1,000 years ago. The ships of the British navy were called the wooden walls of old England. Eight British warships have been called *Royal Oak* since the Restoration in 1660; in 2011 it was calculated that there were also 467 pubs in England called the Royal Oak. Ships made of oak endured. The hardest wood comes from the centre of the trunk and is called heart of oak. This inspired the patriotic song written by David Garrick (music by William Boyce) that was first performed on New Year's Day in 1760, commemorating the 'marvellous year' of 1759, in which the British navy won four sea battles.

> Heart of oak are our ships, heart of oak are our men;
> We always are ready, steady boys, steady!
> We'll fight and we'll conquer again and again!

Oak was used for the most prestigious buildings. It is a pretty straightforward and uncompromising building material, but architects of genius can use it to create surprisingly elegant structures. The hammerbeam technique is perhaps the summit of this craft: short protruding beams that don't meet in the middle, but act as a truss for the roof supports: the Great Hall at Hampton Court Place is a classic example.

Perhaps the greatest oak building of them all was Notre-Dame Cathedral in Paris. Construction began in 1160 and it was finished a mere century later, though there was much alteration in subsequent centuries. The cathedral contained wood from 1,300 trees, which would have covered 52 acres (21 hectares) of land, and the building was sometimes referred to as 'the forest'. In 2019 a fire broke out and a good deal of the building was destroyed; the timber-framed spire collapsed and fell. The limitations of construction in timber were demonstrated plainly enough, but the building certainly had a good long innings. Restoration work began almost at once.

Oaks were so important to German national identity that gold-medal winners at the 1936 Olympic Games in Berlin – 'the Nazi Olympics' – were presented with oak saplings. A British crew won the yachting gold medal in the six-metre class, and the helmsman, Christopher Boardman, ended up with the oak, even though he refused to attend the ceremony. The sapling was planted at his family home at How Hill, in Norfolk. A bomb landed 25 yards away during the Second World War, but the tree survived. It was badly damaged by the gale that hit Britain in 1987, and was pollarded – the top of the tree cut off – in 2013. It's still there, and it's known as Hitler's oak.

TEN

DAISY

Daisies, those pearled Arcturi of the earth,
The constellated flower that never sets

Shelley

The ability to hold two contradictory ideas at the same time is perhaps the defining mark of the human condition: and when it comes to the wild world we show this trait again and again without even being aware that we are doing so. Is it a flower? Or is it a weed? The question is meaningless and yet packed with meaning, revealing important truths, not about plants but ourselves. If you are a botanist, a daisy is a flowering plant. If you are one type of gardener, a daisy is a charming addition to a lawn. If you are another type of gardener, it is an invasive horror that shames the person who perpetrated it and must be wiped out by any means possible.

Daisies have evolved to survive in well-grazed swards. In low grass they put out low, flat rosettes of leaves, and from these, they stick up flowers that are pollinated by insects and then seeds that are dispersed by the wind. They are comfortable on most types of soil. Since humans seek to replicate a well-grazed sward at every opportunity (see Chapter 6), it is inevitable that daisies have followed humans, prospering in the tracks of their mowing machines. Daisies don't do well in long grass, so one solution to daisies, if you consider them a problem, is simply not to cut the grass. Admittedly you will then get plenty of other species coming in, and humans have a great taste for a monoculture.

Daisies survive on lawns because however low you set the blades on your mower, you can never mow a daisy rosette. You can mow the heads, but they will grow again, given half a chance. On uneven ground a daisy flower will often get missed by the blade, bowing under the mower and rising up again after it has gone.

This is troubling to those who believe that a lawn should be 'prestigious'. A grass monoculture that is mowed first one direction and then the other, to create a striped pattern, is a public statement about the owner's status; it frequently means that you can afford to employ someone to do the job for you. Daisies can be eradicated by hand: you can cut them out with a knife at weekly intervals till you have conquered them. You can buy a purpose-built daisy-grubber. But the principal anti-daisy weapon is chemistry: you can spray lawns with selective herbicides; these allow the grass to thrive while killing its competitors.

This is all very strange when looked at objectively. It's partly to do with the human need for control. Cut grass, free from invading plants, is soothing to humans: it shows that nature has been first tamed and then allowed back under licence. We have nature on our own terms. Such management provides a task of perpetual labour, a Forth Bridge job that can never be completed. This too is a soothing idea to many of us. It has to be regular and often: infrequent mowing is actually a help to daisies. But once you have won the battle, you can maintain your advantage over daisies by constant mowing while setting the blades low.

Daisies are good at surviving. If they can survive the steady grazing and heavy footfall of large mammals, they can cope with life on a gently managed lawn. They will flower early and carry on flowering until late: in the UK you can find them blooming in March and in October, even in mild winters. The flowers close up at night and open with the return of the sun, and they also look like small suns. Both the flowers and the sun itself have been referred to as 'the eye of the day'. Chaucer summed it up, putting the daisy in a more favourable light:

> For nothing else, and I shal not lie
> But for to look up the dasie,
> That well by reason men call may
> The dasie or the eye of day,
> The empress and floure of floures all…

Like as many as thirty other species, daisies are useful medically, as a dressing for wounds; another name for daisy is 'woundwort'; the show-stealing villain in Richard Adams's rabbit odyssey *Watership Down* is General Woundwort. Bandages soaked in the juice from daisies were applied to sword and spear wounds after battles; before battles people gathered daisies in sackfuls.

Daisies are also loved for their own sake. They are associated with childhood and innocence. Daisy chains can be made by making a slit in the stem of one flower and inserting the stem of another: and on and on, until you have a crown or necklace of daisies. Botticelli's Flora, in his great work *Primavera*, is wearing daisies in her hair. Daisy is a popular girls' name: the sevententh most popular in England and Wales in 2010, though it had slipped to no. 150 in 2021. The name implies unassuming and unpretentious beauty, beauty of the nicest kind, as reflected in Daisy Buchanan, heroine of F. Scott Fitzgerald's *The Great Gatsby*, and, with rather less ambiguity, in the song 'Daisy Bell (Bicycle Made for Two)', written in 1892 by Harry Dacre and said to have been inspired by Daisy Greville, Countess of Warwick.

Opposite 'There's a daisy': Ophelia, 1894, by John William Waterhouse (1849-1917).

Daisy, Daisy,
Give me your answer, do!
I'm half crazy,
All for the love of you!
It won't be a stylish marriage,
I can't afford a carriage,
But you'll look sweet upon the seat
Of a bicycle built for two!

But Daisyworld was a grimmer matter. This was the name of a 1983 computer simulation by James Lovelock and Andrew Watson. It represented a hypothetical planet with two species of daisies. Its purpose was to validate the Gaia hypotheses of Lovelock, in which he postulates that the earth is a single living organism, a self-regulating and complex system that allows life to sustain itself. The corollary is that messing about with this system is dangerous and destructive.

Daisies were a nice choice for this task. Not only do they capture the simplicity – even naivety – that this deliberately simple computer system encapsulated, but they also reflect our ambivalence about nature. We love daisies, we sing about daisies and we name our daughters Daisy: we can't bear the sight of daisies and associate them with loss of prestige. Do we want daisies in our lives? Or do we not want daisies? Is a daisy a flower or a weed? Only one answer: yes.

ELEVEN

YEAST

Here with Loaf of Bread beneath the bough
A Flask of Wine, a book of verse – and Thou…

The Rubaiyat of Omar Khayyam,
translation by Edward FitzGerald

Yeast is a fungus. Fungi are not plants, but we generally think of them as much the same sort of thing. That's because they are alive like us, but they don't move. We humans prefer a binary view of life – light and dark, good and evil, love and hate – and so we divide the world into animals and plants. But scientific taxonomy – as opposed to folk taxonomy – traditionally divides eukaryotic organisms – those that have cells with a nucleus – into three kingdoms: animals, plants and fungi. It makes people slightly queasy to realise that animals and fungi are more closely related to each other than either is to plants. Animals and fungi are both consumers; directly or indirectly dependent on plants. Plants are the only life forms that are capable of making their own food. I am including nine fungi in this book because fungi matter to us and because fungi are considered plants in folk taxonomies… while remembering all the time that the two groups are radically different.

Where would we be without yeast? We might not have survived as a species – and we might not have formed viable agricultural communities – without yeast. There are around 1,500 species of yeast: single-celled organisms that evolved from multi-celled ancestors. (Note here that evolution does not invariably move towards greater complexity; if simplicity is an advantage, simpler forms will prevail.) The term yeast is informal, and there are species of yeast in two different phyla (a phylum is one level down from kingdom in traditional taxonomy; we belong to the phylum of chordates or back-boned animals).

We normally use 'yeast' for a single species: *Saccharomyces cerevisiae*. This fungus is capable of converting carbohydrates to carbon dioxide and alcohol; the process is called fermentation. Yeast gave us booze and leavened bread. These two things are central to a great deal of human civilisation: so much so that the central ceremony of the Eucharist in Christian churches is based around bread and wine. A visiting Martian anthropologist might conclude that we worshipped yeast as a god.

Yeast is – praise the Lord – available everywhere. Yeast just happens, as we have seen with the making of sourdough bread, described in Chapter 2. It's so easy that

Let the joy begin: The Four Elements of Brewing, *1717, by Giovanni Antonio Pellegrini (1675-1741).*

it must have happened again and again all over the world: leave your dough longer than you meant to and it gathers wild yeast and starts to rise. You cook it anyway, because you can't afford to be wasteful – and your bread is 100 times nicer.

The discovery that yeast makes alcohol would have been, if anything, even more straightforward. Practically anything will turn into alcohol if you give it half a chance: all you have to do is nothing. A woman I know was between jobs, so she worked on a book to be called *God Loves the Drinkers.* Alas she was whirled away into high-powered employment before it could be written, but the truth behind the project remains: alcohol is readily available to us all, rich and poor, and it always has been.

Some of the earliest known writing refers to yeast. The *Hymn to Ninkasi,* the Mesopotamian goddess of beer, contains both prayers to the goddess and a recipe for her greatest gift; it was a useful thing to learn by heart in a civilisation in which not everyone was literate. The written version is about 4,000 years old, the hymn certainly much older. No doubt you sang the hymn as you made the beer:

> *When you pour out the filtered beer of the collector vat*
> *It is like the onrush of Tigris and Euphrates…*

There is evidence for beer from 13,000 years ago. Alcoholic drinks were central to the growth of most human civilisations: for ritual, for ecstatic experiences, for communication with the gods, and also for pleasure and conviviality. Drinking games occur in Mesopotamian legends; in *The Epic of Gilgamesh* of about 1,800 BC, the eponymous hero is encouraged to 'Fill your belly, Day and night make merry…'.

54

Beer is stuffed with calories: offering rehydration and nourishment all in one to people doing manual work; there is evidence of Mesopotamian workers actually being paid in beer. (In Germany today beer is jovially referred to as *flüssiges Brot* – liquid bread.) A fermented drink also keeps better than water; in a warm climate water quickly attracts a culture of undesirable organisms. Though there was no awareness of water-borne diseases – the germ theory of illness was not properly developed until the eighteenth century – bad water smells bad and is unpleasant to drink; you don't need to know about the origins of cholera and typhoid to reject it. In an agricultural society, finding fresh water unpolluted by animal droppings was difficult. To all these problems, beer was a solution. The beer that was drunk for rehydration was often much weaker than the beers we are used to drinking in the twenty-first century. These days beer is reckoned to be the world's third most popular drink – after water and tea. How many deaths has beer prevented across human history? Countless numbers. Would humanity even have survived the prevalence of water-borne disease in agricultural communities, without yeast to provide an alternative to polluted water?

We will look at alcohol further in later chapters, for without frivolity we must accept that it plays a central role in the history of our species. But we must remember that it begins with yeast. When you crush grapes to make a drink, there is yeast already present on the skins: the combination can hardly wait to become wine. Mead is made from honey: no doubt the drink was discovered when people found wild bee nests that had attracted wild yeast and fermented. The Chinese made rice wine, and there were many alcoholic drinks in pre-Columbian America.

We should consider one further use for yeast before we move on, and that is in the manufacture of yeast extract, more usually known by its proprietary names: Marmite in the UK, Vegemite and Promite in Australia, Vitam-R in Germany and Cenovis in Switzerland. It is a useful dietary supplement for vegans, a flavour-enhancer in cooking and a pleasant (or not) addition to bread. In Britain, Marmite has the distinction of being the only food product that has been marketed on the premise that some people loathe it to the point of nausea: a television advert showed a couple on a first date, going back to her flat... but after the first kiss the boy is retching helplessly. The girl has been eating Marmite.

Yeast has given us the essential staple food of many cultures on earth, and also our intoxicant of choice: the source of conviviality, comfort and heightened experience, with a reverse side of belligerence, addiction and destruction. So let us close with a few more lines from *The Rubaiyat of Omar Khayyam*:

Dreaming when Dawn's Left Hand was in the Sky
I heard a Voice within the Tavern cry,
'Awake, my Little ones, and fill the Cup
Before Life's Liquor in its Cup be dry...'

TWELVE

CANNABIS

Take me disappearing through the smoke-rings of my mind…

Bob Dylan, 'Mr Tambourine Man'

All the hope for a new and better society seemed rooted in a plant. There was music in the cafés at night and revolution in the air – and everywhere the whiff of cannabis. The consumption of cannabis was not just about the pursuit of pleasure: it was also about a different way of looking at the world, without ambition, without bellicosity, with kindness and generosity, with deeper perceptions and enhanced creativity. The hippy era has been rightly mocked: can you imagine people ridiculous enough to endorse values like love and peace?

The cannabis plant was an important part of it: its illegal nature creating a shared identity with shared values, which was thrillingly termed 'the counter-culture'. At its coarsest, it endorsed the belief that the world would be an infinitely better place if the world leaders could sit down together and share a good stiff joint. For many of us the interest in the drug faded reasonably quickly, but some of the shared values remained: the idea that people who seek power don't necessarily have the world's best interests at heart, that life is nothing without freedom and that without nature it's not worth living.

Cannabis is still consumed all over the world, though these days seldom part of a programme for world peace. In many places it is now legal. The plant has been in use as a drug for at least 4,000 years, often one with ritual and religious importance.

Three species of cannabis are usually recognised. These produce a substance from which tetrahydrocannabinol, or THC, can be extracted. Why should a plant produce a psychoactive drug? We are looking here for an evolutionary explanation, rather than the notion that God made cannabis so we could all get high. The most likely reason is that the substance protects the plant from the invertebrates who are trying to eat it. Plants evolve self-protective strategies just as animals do: we have already met thorns on roses and will later look at nicotine and caffeine. It's also been suggested that cannabinols protect the plant from microbes that carry disease, and that they protect it from excessive ultraviolet light.

One certainty is that the plant produces cannabinols for its own purposes. But humans somehow learned that their consumption could have startling effects on

'Scuse me while I kiss the sky: image, apparently of Jimi Hendrix and a marijuana plant, by Raymond Lee Warfield.

human brains. No doubt this discovery came from chewing the leaves. The plant had been cultivated as hemp for many years before it was used as a drug: hemp can make a good coarse fabric and, when strands are twisted together, the most excellent rope. Hemp was used by humans for at least 10,000 years, long before we cultivated anything; ropes must have been one of the earliest and most important tools, but unlike stone blades, they don't survive long. Perhaps we should talk about the Hemp Age rather than the Stone Age. Modern industrial hemp is grown from strains very low in THC; it is used in rope, clothing, shoes, as animal feed and biofuel.

Cannabis is indigenous to Central Asia and the Indian subcontinent. It was consumed for religious purposes during the Vedic period of India, between 1500 and 500 BC. The period is named for the Vedas, sacred texts much revered by the hippy movement, if seldom actually read. The earliest evidence of smoking is from the Pamir Mountains in Western China, where the plant was burned on wooden panniers and its inhalation was part of ceremonies in mortuaries 2,500 years ago, perhaps in attempts to contact the dead.

Altered states of consciousness have been part of ritual since religious practices of any kind began. Fasting, rhythmic chanting and dancing, hyperventilating, exhaustion and self-flagellation have all been used, sometimes in combination, to produce an exalted state and the vision of visions. Psychoactive drugs do the same thing easier and quicker.

The consumption of cannabis became part of traditional cultures in places where it grew, and in Western minds was associated with fantasies about the wisdom of the East. The drug became a nineteenth-century exoticism in European culture. Jacques-Joseph Moreau wrote a treatise on the effects of cannabis after travelling in North Africa and Western Asia. This led to the formation of the Club des Hashischins in Paris, which was active 1844-49, and whose members included Victor Hugo, Alexandre Dumas, Charles Baudelaire and Gérard de Nerval. An Irish physician, William O'Shaughnessy, brought a good amount of the stuff to Britain after working for the British East India Company. Texts on the subject were written by Baudelaire, with *Les Paradis Artificiels*, and Fitz Hugh Ludlow, with *The Hasheesh Eater*.

The drug was criminalised in most parts of the developed world towards the end of the nineteenth century. The process began in Mauritius in 1840 and in Singapore in 1870, because it was feared that the drug made the local workers less productive. In 1906 the first restrictions were introduced to the United States.

The drug came into fashion with earlier forms of counter-culture, associated with jazz music and later with the Beat poets, precursors to the hippy movement of the 1960s and its doctrine of peaceful revolution. Revolution of any kind is distasteful to authorities. As a result, cannabis was demonised and popular figures were persecuted. Mick Jagger (now Sir Mick) of the Rolling Stones was sentenced to prison for possession of cannabis in 1967, until there was a rethink occasioned by a famous leader (opinion piece) that appeared in *The Times* under a headline borrowed from Alexander Pope: 'Who breaks a butterfly upon a wheel?' Richard Nixon, while campaigning for re-election as president of the United States in 1972, attempted to have John Lennon deported on the grounds that he had been found in possession of cannabis four years earlier and elsewhere. Lennon had spoken out against the Vietnam War, then at its height, and his song 'Give Peace a Chance' was sung at demonstrations.

The hippy movement faded or took new forms, and cannabis was no longer about revolution but pleasure. Figures from 2016 showed that 51 per cent of adults in the United States had used the drug at some time, 12 per cent in the previous year. The drug, much talked up as a peaceful and intellectual alternative to alcohol, doesn't enhance conviviality, and perhaps for that reason remains less popular. One of the reasons routinely proffered for its illegality is that it is a 'gateway drug', one that, though comparatively safe in itself, leads its users inevitably towards more damaging substances. Alcohol, the world's drug of choice, remains legal in most places, and has damaged far more lives than cannabis. Cannabis, however, is not the invariable bringer of nirvana: heavy and/or vulnerable users have experienced psychosis and other mental problems and been unable to lead normal lives as a result.

There have been recent studies of the medical benefits of cannabis. It is used successfully in pain management, in reducing nausea during chemotherapy, and for increasing appetite. As a result, legal restrictions of the drug have been lifted in many places; as early as 1976 it was legal to smoke cannabis in certain places in Amsterdam. In 2013 the drug was made legal in Uruguay and it is now legal in South Africa, Canada, Georgia, and in eleven of the United States. More modern strains of cannabis, sometimes grown with hydroponics rather than in soil, have been selectively bred for a very high THC content, making the drug a radically different experience from the gentle highs of my youth.

I visited a shop in New Brunswick, Canada, that bore an unambiguous sign reading 'Cannabis'; I went there with a Canadian friend and user. Our ID was checked at the door by a kind lady of a certain age ('It's like scoring from your mom') and then we were served by a bright young lad; it was more like visiting an Apple showroom than buying a quid deal. I felt no urge for reacquaintance with the drug, but noted that Canada had not sunk into degeneracy as a result of this legalisation.

THIRTEEN

ORCHID

*'You won't mind if I put the flowers straight on your bodice; the
jolt has loosened them.'*

Marcel Proust, *Swann in Love*

Flowers are about sex. That's true botanically, and it's also true for humanity, in the way we use flowers as love-gifts and as symbols of love and lust. The Middle English word for orchid is 'bollockwort'. They were called *orchis* by Theophrastus, Greek philosopher, pupil of Aristotle and a great namer of plants; the word means testicle. In both cases it's a reference to the double tuber found in some species.

Orchids are noted for their flowers: for their dramatic and obvious sexuality, for flowers are the plant's reproductive structures. Orchids are also prized for their exoticism, and they have inspired desire and obsession. They have a whiff of excess, of danger. In the nineteenth century, the passion they inspired in many was termed orchidelirium. There are hundreds of societies all over the world devoted to the cultivation of orchids. Charles Darwin wrote to his great friend, the botanist Joseph Hooker: 'I was never more interested in my life in any subject than this of orchids.'

Orchids are a huge family (the *Orchidaceae*) which contains around 28,000 species in 763 genera. They tend to be associated with tropics and hothouses but there are many species in the northern latitudes, fifty-two in the UK, where they attract an obsessive following. In John Fowles's novel *Daniel Martin*, the eponymous hero begins the friendship of his life after discovering an unlikely sharing of 'the orchid mystique'. He finds Anthony painting a man orchid; the petals of the flower resemble a human figure, apparently equipped with knee-length penis.

So showy! Do orchids really have to be so showy? F. Scott Fitzgerald wrote: '"Do you know that lady?" Gatsby indicated a gorgeous, scarcely human orchid of a woman who sat in state under a white-plum tree.' Orchids seem scarcely vegetal: meat-like, flesh-like, animal-like in the desperation of their cravings. Are they so desperate? Yes indeed. They pollinate each other not with individual grains but in large packets called pollinia. These are necessary because of the orchids' strategy for seed dispersal: they produce an astonishing number of astonishingly tiny seeds, which are dispersed by the wind. That requires a great deal of pollen (male) to be attached to the stigma (female) of each flower. Casual dustings are no good: they need a big lump of the stuff or the system won't work.

Never fails: The Sender of the Orchids, 1929, by Clarence F. Underwood (1871-1929).

This seeding strategy gives them two options. The first, the one they use, works on the idea of a great deal of pollen reaching exactly the right flower. The second would be to produce pollen in impossibly enormous quantities. In other words, orchids have no choice but to seek insects, and to attract them in the most dramatic way possible. The blooms are huge and they need to last longer than most flowers. The whole system might look like reckless extravagance, but it's elegantly economical when compared to an impossible pollen super-production. It works well, but inevitably it comes at a cost. It creates a dependency. Many orchids need specialist pollinators: without them, life stops.

A classic example is the bee orchid, which is found in Europe. The flower looks like the female of a species of solitary bee; the orchid also pushes out a scent that mimics that of the bee. A male bee attempts to mate with it, and flies off – only to be deceived all over again by the next bee orchid, which it pollinates. The flower's resemblance to the bee, though remarkable, is by no means perfect, which might encourage people to muse on the bee's stupidity... perhaps forgetting that it's been known for male humans to be sexually aroused by two-dimensional images of women. The species of solitary bee mimicked by the flower is rarely found in northern parts of the plant's range, including the UK. Here plants must rely on the far less effective method of self-pollination. In 1876 Darwin published a work on precisely that subject, contrasting the efficacy of self- and cross-fertilisation. (More on this in Chapter 33 on toadflax.) He demonstrated the weakness of self-fertilised plants, and anticipated some of the discoveries of Gregor Mendel on genetics (see Chapter 4).

In 1863 Darwin had a greenhouse established at Down House in Kent, and he gave a great deal of his time to orchids – 'I long to stock it, just like a schoolboy,' he said as the building neared completion. One of his species was the Madagascan orchid *Angraecum sesquipedale*, on which he measured a nectary almost a foot (30cm) in length. How did that ever get pollinated? 'The pollinia would not be withdrawn until some huge moth, with a wonderfully long proboscis, tried to drain the last drop,' Darwin suggested. This speculation was the subject of ridicule... but forty years later a sub-species of hawkmoth with just such a proboscis was discovered: it was duly named *Xanthopan morganii praedicta*: the told-you-so moth. It is a perfect example of the principle of coevolution: the way the development of one species reciprocally affects the development of another. Darwin published *Fertilisation of Orchids* in 1862, three years after *On the Origin of Species*; you could look on it as a told-you-so book.

But it was not the lust for pure knowledge that propelled the orchid boom in Europe. It was more the lust for pure lust: sex, money, power, all the usual stuff. Tropical orchids are fabulously showy, and in the nineteenth century they were hard to obtain. You could only get them if you were very rich, so wearing an orchid

Author as dandy: detail of the Portrait of Marcel Proust, *1892, by Jacques-Émile Blanche (1861-1942).*

in your buttonhole or your bosom was a powerful statement. There is a portrait by Jacques-Émile Blanche of Marcel Proust, at the age of twenty-one, an exquisite, a writer with a slight but pleasing talent. In the picture he contrives to look both uneasy and self-satisfied at the same time as he gazes back at the viewer with a cattleya orchid – a particularly showy genus – in his lapel. Years later, in the early part of his great work À *la Recherche du Temps Perdu*, he writes of Charles Swann's pursuit of the courtesan Odette. Swann makes his move at last in a carriage after the horse spooked and disturbed both the lady and her corsage: 'Look, there is a little – I think it must be pollens, spilt over your dress, – may I brush it off with my hand?'

He does so, one thing leads to another: and afterwards the act of love is known between them as 'doing a cattleya'.

Orchids were big business. Orchid collecting was an adventurous way to seek a fortune. In London, John Day of Tottenham established a big collection and business based on orchids. Sir Trevor Lawrence paid £235 (perhaps £100,000 today) for a single specimen of what became known as *Aerides lawrenciniae. The Boys' Own Paper* (founded in England in 1879) ran a series of stories called *The Orchid Hunters*.

One of the great attractions and mysteries of orchids was the fact that they are extraordinarily difficult to reproduce. For most species it couldn't be done at all. The problem is that the seeds will only germinate when they have access to nutrients that come from the fungal threads called mycelium. In 1922 a way of injecting the sowing medium with the right sort of nutrients was discovered and from that point on, orchids became democratised.

They maintained their reputation for exoticism. Rex Stout's detective Nero Wolfe (created in 1934) was devoted to the study of orchids; in Raymond Chandler's *The Big Sleep* (published in 1939), Philip Marlowe meets General Sternwood in his orchid house where the stalks of the plants were 'like the newly washed fingers of dead men'. But these days you can buy orchids from the supermarket when buying a tin of beans. There is a huge illegal trade in wild orchids, which does great ecological damage, so it's good to check the provenance before you buy. But it's also easy enough to buy orchids legally: mystery and exoticism for all. Or is that a contradiction? Can an orchid ever be really homely?

FOURTEEN
BRAZIL NUT TREE

'I'm Charley's aunt from Brazil – where the nuts come from.'

Brandon Thomas, *Charley's Aunt*

Once we called it jungle and understood it as a place where nature ruled and only the bravest men dared enter. We made it a place of fantasy: Jungle Comics, which began publishing in 1940, routinely showed a white man in leopard-skin shorts wrestling with crocodiles and gorillas. When we talked about the Law of the Jungle we meant values infinitely more ferocious than those of the civilised world. The jungle was useless but gorgeously wild: to be tamed in the knowledge that ultimate taming was impossible. Now we call it rainforest, and it is an emblem of fragility. Some see it as a vanishing world of impossible beauty.

We find it hard to see the trees for the wood. It is the ecosystem itself that has us enthralled. But a few individual species stand out – and do so quite literally, for lowland wet tropical forest is characterised by a closed canopy from which, every now and then, a still greater tree stands taller than the rest. These trees are unromantically termed emergents, and the Brazil nut tree is a classic example. Its trunk seldom exceeds a couple of metres (6ft 6in) in diameter, but it can soar to 50 metres (160 feet).

We are all familiar with the seeds, which we consume and call Brazil nuts, though Brazilians more often call them *castanhas-do-para*, 'chestnuts from Para'. There is a beautiful truth about these excellent fatty nuts: they can't be cultivated. They can only be gathered. They can only reach our tables by way of undisturbed forest. Attempts at establishing Brazil nut plantations have routinely failed. Every time you eat a Brazil nut you are celebrating the perfection of intact forest. If they are to produce fruit, the trees must be pollinated, and their semi-enclosed flowers can only be pollinated by species of large-bodied bees that can't survive away from undisturbed forest.

The fruit take fourteen months to develop, but trees like this can afford to take their time; a Brazil nut tree can easily live for half a millennium, and twice that age is far from unknown. The fruits weigh a couple of kilos (more than 4lbs), and when at last they are ripe, they come crashing to the ground. They look a bit like the inside hairy shells of coconuts. You find the seeds – the Brazil nuts – packed inside like segments of an orange, up to twenty-four in a fruit.

Forest giant: eighteenth-century watercolour drawing of a Brazil nut plant by José Codina.

The fruit have a hole in them: a helpful starting point for agoutis, the main predators of Brazil nuts. They are rodents and look like long-legged guinea pigs. This helpfulness is not a mistake. They can't eat all the seeds at a sitting, so they bury the rest and come back later. Some seeds will survive uneaten. These eventually germinate and the resulting saplings remain more or less dormant in the darkness of the forest floor, where only 2 per cent of sunlight ever manages to penetrate. But when a storm takes down one of the tall trees, an opportunity is created: a tsunami of light. The Brazil nut saplings (and their competitors) then strive for the canopy… and those that win the lottery reach it and emerge beyond. It sounds a chancy way to live, but over the course of centuries, long shots become certainties – so long as the environment remains undisturbed.

The fallen nuts are gathered by humans as well as agoutis. In some places this is a community operation, in others a business employing itinerant labour. The future of the industry depends on restraint: in places where the nuts have been gathered intensively there is no regeneration: the absence of young trees makes for a finite industry. Where the gathering has been light, the presence of young trees indicates sustainable practice.

The nuts are prized because they contain selenium, which is helpful to the immune system. Brazil nuts have the distinction of being the only radioactive food routinely taken by humans: Brazil nuts contain radium, though as most of us can attest, it comes in sub-lethal quantities. The oil from Brazil nuts is useful in clocks, and it is also used in cosmetics and paint. The wood is of high quality for construction and flooring, but its importance as a producer of nuts gives the Brazil nut tree a better chance than most of being passed over.

The destruction of rainforest has been the story of the last 100 years. It was seen first as a benign act: taming the unspeakable to create a more civilised planet. Now, in a classic piece of ecological revisionism, it is widely understood as a self-willed disaster: one that began in ignorance and now continues in full knowledge of the terrible consequences. Briefly, rainforests were the most productive places on land for taking in carbon dioxide and giving out oxygen. Carbon dioxide is a greenhouse gas, and its increasing presence in the atmosphere means that the planet keeps more of its own heat than it used to. Global temperatures are measurably rising, putting the world out of ecological balance. We are already 0.8°C warmer than at 1900, and by the end of the century the rise has been calculated at a further 2.6-4.8°C. Rainforests also transpire water, which creates clouds and rainfall. By destroying rainforests we are losing a system that generates oxygen and water.

A great deal of the destruction has taken place under a false premise: that rainforests are gloriously fecund because they stand on gloriously fecund soil. It's not true. Certainly rainforests have great access to warmth and moisture, and have done so for more than 50 million years. This combination of richness and stability has created an almost impossibly complex system of inter-dependencies. But the fertility of rainforest is not in the soil; it is in the forest itself: in the organic matter that reaches the forest floor and is then recycled. But still the forests are destroyed, against the best interest of the species that is doing the destroying. To eat a Brazil nut is to understand, for a fleeting moment, not with your intellect but with your gut, that rainforest brings us benefits we cannot even begin to seek elsewhere.

FIFTEEN

OIL PALM

I scream, you scream, we all scream for ice cream.

Howard Johnson, Billy Moll and Robert A. King

How much palm oil did you get through last year? In 2015 everyone on the planet consumed on average 17lbs (7.7kg) of the stuff in the course of the year. The more processed and packaged food you eat, the more palm oil you use. It's in bread, crisps, margarine, ice cream, pizza, instant noodles, chocolate and many other foods. It's also in soap, shampoo, lipstick and toothpaste. There's oil palm oil in 50 per cent of all packaged foods. Palm oil is so useful it seems almost magical. It's increasingly useful in biofuels; we'll look at that issue in more detail in Chapter 95 on oilseed rape, noting in passing that biodiesel from palm oil has three times the carbon emissions of fossil fuels.

Palm oil has been used for cooking for 5,000 years; several litres were found in vessels in Abydos, one of the oldest cities of Ancient Egypt. There are three main species of oil palm tree; the American, the maripa oil palm of south America and the African, by far the most important plant for cultivation. When the industrial revolution came along in the eighteenth and nineteenth centuries, palm oil was used as a lubricant, and also for the manufacture of candles; much of human civilisation is about the fight against darkness. By 1870 palm oil was the principal export of a number of West African countries until it was supplanted by cocoa (Chapter 72). In 1910 the plant was introduced to Malaysia, and in Southeast Asia the plant found its true home. It's possible that the oil palm was first grown there for ornament; they are compact but pleasing trees.

The oil was useful from the start, but nothing like as useful as it became in the two decades after 1940, when food preparation became an industrial process, rather than an agricultural and domestic one. When you could buy a cake rather than bake it, our understanding of food – and by extension the world – changed radically. Palm oil was a crucial part of that shift.

You get the oil from the fruit and the kernel. It has an exceptionally high content of saturated fats. That might not seem such a marvellous thing, but it's the reason for the plant's success. Saturated fats are not nearly as damaging as trans fats. These come from meat and dairy – but also, and more dangerously, from hydrogenated vegetable oils, treated to give them a longer shelf life. Food

containing such fats are now banned in the United States and elsewhere, because of the health risks they pose: bad cholesterol leading to heart disease and stroke. At one stage all fat was demonised and a fat-free diet was a great goal. It was subsequently found that certain kinds of fat – monounsaturated and polyunsaturated – are actively good for you. They can be found in nuts, especially Brazil nuts (see previous chapter), and also most famously in olive oil (more on olives in Chapter 82).

The saturated fats found in palm oil are in between the good and the bad fats and are therefore considered acceptable in many processed foods and cosmetics. Saturated fats are more or less solid at room temperature: they include butter and cheese – and palm oil. Palm oil was used for soap in the United States under the brand name Palmolive, and in Britain as Sunlight Soap, which was the world's first packaged and branded laundry soap. It was marketed by Lever Brothers from 1884 and is commemorated in this ancient pastiche of a Christmas carol:

> *While shepherds washed their socks by night*
> *All seated round the tub,*
> *A bar of Sunlight soap came down*
> *And they began to scrub.*

The tree produces bunches of fruit. The oil can be refined, bleached, deodorised and processed to become still higher in fat. This has created an enabling substance, one that allows processed food to happen. It's not only versatile, it's cheap: much cheaper than butter, which involves the participation of cows, and margarine, which in the 1950s and 1960s used blubber from whales. The waste product of crushed kernels is used as animal feed.

Palm oil is increasingly useful. We all consume more of the stuff than we did in 2015. To get the same quantity of oil from sunflowers (Chapter 8) you would need ten times the acreage used for oil palms. The oil palm industry has created a dependency, and, as the International Union for the Conservation of Nature observed, it is here to stay. The stuff is inescapable.

It grows best in tropical climates and 85 per cent of the world's supply now comes from Malaysia and Indonesia. The oil palm heartland is the island of Borneo: three quarters Indonesian and the rest Malaysian, apart from the small kingdom of Brunei. The industry here has plenty of problems. There are consistent accusations that parts of the industry use child labour, and that workers are kept in conditions approaching slavery. Some of the work in Malaysia is done by illegal immigrants from the Indonesian side. Most oil palm plantations have been created by destroying rainforest.

Opposite Ubiquitous. oil palms at Tijuca, Brazil.

End product: Sunlight Soap packet.

When I went to Borneo to work with local conservation organisations, the first thing I was told was surprising: it's not us good guys against the bad guys of palm oil. Cynthia Ong, director of LEAP, said: 'This is not a polarisation. It's not about palm oil versus conservation, or palm oil versus forests or palm oil versus orang-utans. That's not going to help anybody. It's about working with the situation in front of us.' Conservationists must work with the industry, and some sections of the industry want to be seen as – and perhaps even want to be – responsible and sustainable.

The contrast between the neat monoculture of the oil palm plantations and the mad diversity of the rainforest is offensively stark. From the air it looks unbearable. The stars of the Borneo forests are orang-utans and pygmy elephants: the continuing expansion of the oil palm industry compromises their future. With the ever-increasing human population – growing by an estimated 82 million a year as I write – we are running out of room for anything other than human-run monocultures. And that comes at a price.

SIXTEEN

CALAMITES

'Oh where are you going to, all you Big Steamers,
With England's coal, up and down the salt seas?'

Rudyard Kipling

Calamites are extinct plants. They come from the group known as horsetails. One or two species still survive, but once this group was dominant. Calamites could grow 30 metres (100 feet) tall. Had they not existed, the course of human history would have been very different. They – quite literally – fuelled our shift from a mostly agricultural species to a mostly industrial one. Calamites grew as the understory in the great wet forests of the Carboniferous era. They lived and thrived and died, and when they did so, many of them went on to form peat; peat is formed when plants decompose in acidic conditions. When peat is subjected to huge pressures from rock laid down in subsequent millennia, it forms coal – and coal was the great game-changer. Coal powered the Industrial Revolution; we now know that coal is also the main driver of climate change.

Coal burns. The process of burning releases energy in the form of heat. Mastery of fire is one of the very few hard-and-fast distinctions between human and non-human animals (see Chapter 22), and the knowledge that coal burns is an ancient part of human history. In some places coal is or was easily accessible: at or very near the surface. Discovering that it burns like wood, only better – hotter and longer – is no great mastery. There were coal fires in China a good 6,000 years ago. Coal was known in Europe in the ancient world: Theophrastus, born in the fourth century BC, noted that 'those known as coals are made of earth, and once set on fire, they burn like charcoal'. The Aztecs used coal in Mexico. Coal is mentioned in the *Anglo-Saxon Chronicle* of 852.

But all these were affairs of outposts. It was in the eighteenth century that things started to change. Advances in techniques for obtaining coal from the ground – mining rather than scratching the surface – coincided with stupendous advances in technology. The most significant was the discovery of steam power, which can be harnessed to perform tasks that were previously impossible or recklessly expensive of humans and other resources.

James Watt didn't invent the principle, despite the legend of the boiling kettle with the bumping lid: 'If we could only harness that power…' Thomas Newcomen invented the first steam engine in 1712, but the engine Watt came up with in 1776 was a great deal more efficient. His genius was not in discovering the principle, but in exploiting it. Now the energy generated by the engine could be converted into rotary motion: round and round instead of merely up and down: not just pistons but wheels. Humanity was off and rolling into a new age. The power of steam – the power of burning coal – could be used for purposes undreamed of before.

These lumps of black earth, created by brutal geological forces over the course of Deep Time and made from the Carboniferous forests and the calamites that grew in it, now changed the way humans lived. The process began with textiles, the first industry to become fully industrialised, and it has been going on ever since. Most of the things we have in homes, our workplaces and in between them are not made but manufactured: created by industrial processes.

Fuel of the future: Calamites plants in a reconstruction of a Carboniferous-Permian river delta, c.360-350 million years ago.

Coal was at the heart of it all, and even today, when we are usually reckoned to be past the peak of global coal use, coal provides the energy we crave, on which we have become dependent. The problem lay in getting hold of the stuff: because we needed more and more of it. The success of the Industrial Revolution drove the increase in population, and teeming humanity drove the industries.

The coal burned; the steam turned the wheels. Populations grew and cities expanded to fit them: and in winter the people kept themselves warm by burning coal. The atmosphere was filled with smoke. That still has good meanings for us: the chimney pouring out smoke tells us of a warm welcome within; a speeding train with black smoke streaming back over the following carriages gives us romantic feelings about travel. Coal was common and coal was cheap, mostly because the labour that fetched it from the earth was cheap. The obvious solution to all the problems of the world was more and more coal.

But the outpouring of coal poisoned the atmosphere. It gave rise to thick, choking fogs that would sit over cities on windless winter days. These were the notorious pea-soupers of London, famously associated with the Sherlock Holmes stories of Arthur Conan Doyle, though he never used the term. Such a fog was also known as a London particular, a mixture of smoke, soot and sulphur dioxide, which gave the fog its sinister colour:

> The yellow fog that rubs its back upon the window-panes,
> The yellow smoke that rubs its muzzle on the window-panes,
> Licked its tongue into the corners of the evening…

This from T. S. Eliot's 'The Love Song of J. Alfred Prufrock': a London fog or smog if ever there was one. (Smog is a contraction of smoke and fog.) In 1952 the Great Smog of London lasted for six days, killed 4,000 people, made 100,000 ill and caused great disruption to transport and industry. It eventually prompted the Clean Air Act of 1956, in which there were restrictions on the use of coal, especially domestically. All the same, there was a similar event in 1962; I remember going to school in fog and trying to see if I could see the hand in front of my face. The answer was: just about.

Coal fuelled the Victorian notion of romantic success: but led to a feeling that industrialisation was not an unmixed blessing. D. H. Lawrence's *Lady Chatterley's Lover* is not just about sex: it's also about coal. The joys of Lady C and the gamekeeper are set against the misery of the industrialised nation; the impotent and crippled husband Sir Clifford Chatterley owns coalmines, seeks to modernise them and has no feelings for nature. Lawrence writes of the coal-miners: 'Incarnate ugliness, and yet alive! What would become of them all? Perhaps with the passing of the coal they would disappear again, off the face of the earth.' George Orwell wrote about the way we get coal from the earth in *The Road to Wigan Pier* of 1937, a book that shocked with its calm, accurate reporting.

In the following years, more problems with the burning of coal were discovered. Acid rain is caused when water mixes with emissions, mostly from coal burning, of sulphur dioxide and nitrogen oxides. (I used to carry an umbrella with the words 'Stop Acid Rain'.) The burning of coal is associated with climate change, for it releases carbon dioxide into the atmosphere, and also methane, which is twenty-one times more effective as a greenhouse gas. Coal was still burned to generate electricity: some Londoners remember the smoking towers of power stations at Battersea, at Lots Road in Wandsworth, Bankside Power station that became the Tate Modern art gallery, and many others.

But now coal is being phased out in most of the world. That has given rise to a new form of reckoning: peak coal. This represents the year in which a nation measurably began to decrease its use of coal. Germany reached that point in 1985; the United States in 2008; in China they are still building new coal-fired power stations. In many places other sources of power are used instead wherever possible: relatively clean fossil fuels like oil and natural gas, the expensive and dangerous technologies of nuclear power, and the so-called renewables: water, sunlight and wind. It's been calculated that our coal stocks will take us to 2060; their actual use would lead to a still greater rise in global temperatures than those already foreseen.

Coal – decomposed calamites and other plant species – changed the planet irreversibly. These plants changed the way humans live, and have played a major role in what Sir David Attenborough has called the greatest crisis that humanity has ever faced. By this, he meant not the Black Death or world war, but climate change.

SEVENTEEN
RICE

girls planting paddy
only their song
free of mud

Konishi Raizan

For people who live in the West, rice is a kind of food. But for two-fifths of the world rice is food itself. When you eat food, you eat rice. It's the heart of everything – the soul too, probably. If you are speaking Cantonese, you don't say 'let's eat'. You say '*sik fan aa!*' – 'let's eat rice'. Eating rice *is* eating. You might have congee – rice porridge – for breakfast, dim sum at lunchtime, which comes with plenty of rice, and supper at a crowded round table filled with shared dishes – and rice. Lots of rice. You may be wealthy, you may be aspiring middle class, you may do a lower-paid job, you may live in poverty: but you still eat rice, and you eat rice not with but *for* every meal. I travelled the Kowloon–Canton Railway routinely when Hong Kong was still British, and every returning Chinese family carried an electric rice cooker bought in Hong Kong. Not an exaggeration: it was everybody's priority purchase.

Rice is a kind of grass seed. Two species have been widely domesticated: one from Africa, one from Asia. Rice has the third largest global production in terms of crop tonnage, after maize and sugarcane; the last two are also grown to feed livestock. Rice is the most important grain that is grown for humans, more important even than wheat. It accounts for one fifth of all the calories consumed by humankind across the world.

Rice was first grown in the Yangtze River basin in China: some claim for more than 13,000 years. There is a tale about that: disastrous floods had wiped out all the crops and the people were in despair. A dog appeared with seeds adhering to his tail. They planted these – and discovered rice. Rice has been cultivated in India for about 7,000 years and in Africa for about 3,500. It was known in Ancient Greece and Rome; returning soldiers from Alexander the Great's conquests brought rice to Europe in the fourth century BC. Rice travelled to the Americas with the European colonists 500 years ago.

Rice is normally grown as an annual: planted fresh every year. But it can be grown from its own roots in favourable conditions in the tropics: what's called a

ratoon crop. You harvest it by cutting, leaving the roots to recover and they produce new shoots: a system that can work for as long as thirty years on the same roots. Rice does well in areas of high rainfall, and has comparatively low labour costs.

Rice is traditionally grown in paddy fields: you flood the field after or during the setting of the seedlings, to a depth of 2-4 inches (5-10cm). You germinate the seedlings elsewhere, introducing them to the mud at the appropriate moment. This process has a triple advantage: it gives the thirsty crop constant access to water, it suppresses plants that might compete with the rice, and it makes it harder for animals trying to feed on the growing rice. This system requires sound management: no good trying to manage a paddy field if you can't stop the leaks. You can grow rice in this way while keeping ducks on the paddies, which feed on marauding invertebrates: a good integrated system.

The cultivation of rice in paddy fields has inspired one of the world's most spectacular pieces of primitive engineering. In Banaue in Northern Luzon, in the Philippines, the rice paddies stretch up entire mountains, turning each one into a series of giant steps. The sight from a distance baffles the mind; you can walk these hills along narrow walls with tiny wet fields or pools either side – and always in your ears the tinkling of the water, falling from the forested mountaintops through the exquisitely engineered irrigation system until it reaches the bottom. The terraces are reckoned to be 2,000 years old, made with the most basic tools. The fields cover 4,000 square miles (more than 10,000 square kilometres), and in length add up to 12,500 miles (20,000 kilometres) – about half the circumference of the earth. They are a landscape-scale installation celebrating the mountainous importance of rice.

It is not essential to flood the fields to grow rice, but dry cultivation makes the crop more labour-intensive, since it requires more weeding. Once harvested, the seeds are milled to remove the outer husks, that gives you brown rice. This is usually put through a further process, removing the bran and the germ to create white rice. White rice keeps better, but has fewer nutrients. There was a vogue for brown rice during the hippy era: it was reckoned to be a wonder-food that the conventional world had foolishly let go. It was celebrated by Bob Dylan on his 1964 album *Bringing It All Back Home*, complaining about the woman who answers his craving for food with 'brown rice, seaweed and a dirty hot dog'. In sober fact, eating white rice can give rise to beriberi, which is caused by lack of thiamine. Brown rice has ten times the amount of thiamine (vitamin B1), and is also full of good fibre.

Rice must be boiled; it needs to absorb liquid before it becomes palatable. It can be fried first, which adds flavour to the dishes of many cultures, which include

Opposite *Watering the plants: rice-planting in a summer thunderstorm, 1857, by Ando Hiroshige (1797-1858).*

risotto, pilaf and biryani. Rice is important in places that don't require rice at every meal, but not every culture is comfortable with it. Uncle Ben's Rice is sold in packets already parboiled, a process that keeps in some nutrients and is resistant to weevils. It was the top-selling brand of rice in the United States between the 1950s and the 1990s. The packet shows the face of a white-haired African American: senior slaves or servants were traditionally given the courtesy title of 'uncle'.

Most cultures have some form of rice pudding, generally cooked with milk and sugar. It is traditionally associated with children and invalids: a nutritious and easily digested food. It is said that Buddha's last meal before his enlightenment was rice pudding; it was the end of his period of extreme asceticism and the beginning of his journey along the Middle Way (see Chapter 30 on the Bodhi tree). In English tradition rice pudding is associated with dreary wholesomeness. A. A. Milne, creator of Winnie-the-Pooh, wrote a poem called 'Rice Pudding', which begins:

What is the matter with Mary Jane?
She's crying with all her might and main,
And she won't eat her dinner – rice pudding again –
What is the matter with Mary Jane?

In the so-called Green Revolution of the 1960s there were attempts to create what was known as miracle rice: disease-resistant, shorter and stockier plants that avoid loss from drooping, and with a much higher yield. For various reasons, including poor soil quality, it didn't work as hoped.

A problem with modern rice growing is that rice paddies emit methane. The oxygen-free conditions created by the flooding create a soil in which certain bacteria thrive, and these give out methane. Methane is the most potent of the greenhouse gases that contribute to global heating; 1.5 per cent of the world's methane is generated by rice production.

Rice naturally and inevitably turns into alcohol. Rice wines are part of rice cultures. China produces rice wine, and Japan does so in the form of sake. Rice is also used in the making of many beers. Barley (see Chapter 29) is the traditional grain for beer, but maize (see Chapter 94) and rice can also be used, usually alongside barley as what the industry calls 'adjuncts'. Beers with a good deal of rice tend to taste clean, uncomplicated and refreshing; these include Japanese brands like Sapporo and Asahi. Budweiser beer uses rice as well as barley. Rice flour is also used in gluten-free flours, which tend to contain a mixture of different grains.

But such things are by the by. For many people rice is more or less synonymous with breath. Rice is more, much more, than bread is to a Westerner: not so much the stuff of life as life itself, the stuff that allows us to get to tomorrow.

EIGHTEEN

BINDWEED

Long live the weeds and the wilderness yet.

Gerard Manley Hopkins

Bindweed is a plant of striking beauty. It produces flowers in white and pink trumpets in immense profusion: seen, perhaps from a train window, a chain-link fence transformed into a wall of flowers. But the plant is widely hated. Bindweed is remarkably proficient at staying alive, growing and spreading: aims much pursued by humanity. But bindweed is inconvenient to humans, twining around plants that we have grown for food or for pleasure, taking more than their fair share of space, sunlight, rain and nutrients. Once established, it's extremely hard to get rid of.

The course of history changed with the invention of agriculture 12,000 years ago. Once we had agriculture, we could have settlements, stability, co-operation, neighbours, warfare and ruling classes… and it all comes down to plant choice. Human civilisation began with one decision: let's make this plant grow here, but not that plant. This plant is good, all the other plants are bad. We traditionally see life as a series of dualities – work and play, love and hate and so on – and by far the most important of these is good and evil. Perhaps this notion got into human minds by way of agriculture: from the simple understanding that wheat is good and competing plants are evil.

That view is made explicit in St Matthew's Gospel, in which Jesus tells of a man's crop that had been polluted by the sowing of tares, or weeds. Don't try to solve the problem now, the owner tells his servants – wait till the harvest. 'Gather ye together first the tares, and bind them in bundles to burn them: but gather the wheat into my barn.' Jesus explains: 'The field is the world; the good seed are the children of the kingdom; but the tares are the children of the wicked one; the enemy that sowed them is the devil; the harvest is the end of the world; and the reapers are the angels.'

Wheat for angels, tares for the devil. The parable is powerful because it is blindingly obvious that wheat is good and tares are bad. In the original Greek of the gospels, the word translated as tares (in the King James Version of the Bible) is *zizania*, which has later been translated as darnel, a species of ryegrass that looks like wheat when both plants are young.

Bindweed belle: Fairy Resting on a Leaf, c.1860, by John Simmons (1823-76).

We have a word for a plant that gets in our way: weed. In a garden, children ask the hoe-wielding adult: 'Is that a flower or a weed?' It can only be one or the other. A plant is either meant to be there or not meant to be there; if it's not meant to be there it's a weed and a thoroughly bad thing. In France, weeds are *mauvais herbes*, naughty herbs. Eradicating weeds is, then, a work of righteousness.

The ability to deal with weeds is at base a life-and-death issue, not just for the plants, but for those who seek to control them. When humans invented agriculture, the future of the community depended on its ability to encourage the chosen plants and to discourage their competitors. There is a vocabulary of pejorative words to deal with unwanted species: terms that imply a moral failure on the part of the plant.

And also on the person looking after the land. One of the pleasures of growing vegetables on an allotment is discussion of the shortcomings of absent neighbours: to look on a poorly managed plot festooned with the pink trumpets of bindweed: 'Why do they have 'em when they can't look after 'em?'

Bindweed is just one of a number of tenacious plants that threaten productiveness and profitability of commercial crops. A list of principal plants that trouble farmers in the UK might include cleavers, speedwell, mayweed, groundsel, charlock, cranesbill, chickweed, field pansy, sow thistle, fat hen, shepherd's purse, creeping thistle, fool's parsley, dock, knotgrass, poppy – and bindweed.

Hedge bindweed is native to Europe and North America; field bindweed was accidentally exported to America in the eighteenth century in supplies of grain. It remains one of the most serious of agricultural pest plants: creeping above the surface, climbing the stems of plants and weighing them down. It grows very fast in favourable conditions, often outpacing the plants that humans are growing for themselves.

What you see on the surface is not the worst of it: the real problem is the massive and complex root system. It's hard to get accurate information on this, because we have a profound need to exaggerate the hostility of nature, choking the life out of our food and ourselves. Bindweed roots can grow 9 metres (30 feet) deep in the soil; up to 30 metres (nearly 100 feet) has been claimed. New plants can sprout from seeds twenty years old; the plant will also regenerate from severed roots, so breaking up the roots is no good.

You will never be short of suggestions about getting rid of the stuff. Plough every three weeks for seven years: that ought to do it. Or try mulching: placing stuff on the surface that blocks the light, denying the plant what it needs to live. Five years of darkness should kill off the bindweed.

There are less extreme measures. In a garden, you can mulch for six months, uncover it and then fork the soil. After that you pick out all the visible stems of bindweed. These labour-intensive methods work in a small area, but can be dangerous; you can easily damage the plants you are trying to save. Apparently three years of 'ruthless' weeding will get rid of the plant. 'Be persistent. Don't let it flower.'

But the answer more often chosen is chemical. Again you have the problem that you might damage the plants you are seeking to protect. A gel applied directly to the bindweed will do the job. The use of glyphosate is effective on bindweed and many other plants: it is a broad-spectrum herbicide. We prefer to call it weedkiller, as if it only killed the evil plants. Glyphosate was first created by Monsanto in 1970; Monsanto, the American company involved in agricultural chemicals and biotechnologies, was founded in 1901 and sold to Bayer in 2018. Glyphosate is in widespread use across the world, most often under the Monsanto brand name of Roundup.

There are problems with the use of glyphosate, apart from its indiscriminate nature. It gets into food; the World Health Organization said in 2015 that it was 'probably carcinogenic in humans'. Strains of plants resistant to glyphosate are beginning to emerge. In North America its reckless use has affected the growth of wild milkweed, the food-plant of the caterpillar of the monarch butterfly. The monarch is famous for its spectacular migration and is now in steep decline. Glyphosate is now banned in some countries.

The solutions to the problem of bindweed and other unwanted species are as much a problem as the initial problem. Humans have been caught up in an arms race with certain species of plant for 12,000 years: and with rising human populations there is an increasing demand for food, and therefore for technologies that increase yields. We have chosen which plants are good and which are not. The effects of those choices are widespread and complex: by no means unambiguously good or, for that matter, unambiguously evil.

NINETEEN
FIELD POPPY

What passing-bells for those who die as cattle?
Only the monstrous anger of the guns.

Wilfred Owen

The field poppy – a plant with bright red ragged flowers – was for many years a familiar sight in cornfields. Once established it couldn't be removed without risking damage to the crops, so as the wheat turned from green to gold the exultant red of the poppies added to the glory of the landscape. I once played cricket on a ground with a thatched pavilion, shire horses in a field along one side of the boundary and on the other, a wheatfield aflame with poppies: a timeless ritual of twisters and lobs and draw-shots and long-stops while the church clock stood forever at ten to three.

So, no, people are not normal about poppies. They are not your normal weeds of cultivated land. They annoy farmers and send a gaudy banner across the countryside about his failure to maintain a strict monoculture: but non-farmers who pass by feel a stab of nostalgia, responding not just to their beauty but to vivid memories of times they never knew.

Poppies have certain advantages when it comes to involvement with agriculture. Their seeds are long-lived, staying viable in the soil for twenty years, and they will germinate when the soil is disturbed – and since cultivation involves disturbing the soil, poppies are in a strong position. The plants grow flowers and disperse seeds before the wheat is harvested; by the time the crop is gathered in the seeds are set for the following year. They are spread by the wind: when the petals have dropped, the seed heads remain, swaying with the wind and scattering seeds. The plants flower throughout spring and summer, and such lack of fussiness is always an advantage in environments controlled by humans. Poppies are also mildly poisonous to grazing animals, so they resist being eaten.

Poppies were common in Europe and most of Asia east to Pakistan, as far north as Scandinavia and the Baltic states. These days they are seldom seen in cultivated fields: the use of herbicides, most often glyphosates, as discussed in the previous chapter, has more or less wiped out the poppy-strewn cornfield. Agriculture is more efficient at producing food; the countryside is less efficient at pleasing human senses. Poppies turn up on roadside verges in Britain in the places where the local council is not too crazy about mowing.

Wild poppies have been given various popular names to distinguish them from other sorts of poppies (see Chapter 57 on the opium poppy): field poppy, common poppy, corn poppy, corn rose, red poppy and Flanders poppy. Poppies have been domesticated as garden plants, and cultivars produce flowers in startling colours: yellow, orange, pink and white. But it is their redness that really counts, especially in Britain: that and their Flanderness, as it were.

Poppies tell us about the power of symbols and their shifting, slippery nature. There are times in Britain when the wearing or not wearing of a poppy can seriously affect a person's career, and might cause the fall of a Prime Minister. They have become a symbol of British patriotism, but not of a straightforward kind. You have to wear one to show that you care about the sacrifices made by members of the British armed forces.

Sentimentalising war: In Flanders Fields, c.1919, *by Willy Werner (1868-1931).*

Poppies, as we have seen, thrive in disturbed ground. It follows that when the ground was disturbed as never before in the First World War, poppies sprang up like anything. They grew between the lines of trenches, poppies bright and vivid in No Man's Land. The Western Front was in two provinces of Belgium and one of France, taking in Passchendaele and Ypres, an area collectively known as Flanders. We mostly think of this area – campaigned over across the centuries – as Flanders Fields; we do so because of a poem, 'In Flanders Fields', by a Canadian physician, Lt-Col John McCrae. This simple eloquent poem established itself in the minds of generations, most of them unfamiliar with the poem itself:

In Flanders Field the poppies grow
Between the crosses, row on row…

The poem was written after the Second Battle of Ypres, and was published in *Punch* in 1915. It takes the form of an explicit message from the dead: keep on fighting. There is nothing of the anguish and despair of the truly great poets who also took part in the First World War, Siegfried Sassoon and Wilfred Owen: war must continue, or the dead died for nothing. The equation with the poppies, the colour of blood, and the war-dead is here established for all time.

This idea was taken up a Frenchwoman, born Anna Boulle in 1878. She lectured in England and later in the United States for the French cultural organisation Alliance Française, but she is remembered, by her married name Madame Guérin, as an indefatigable fundraiser. She had the brilliant idea of centring the fundraising for widows, orphans and veterans on Armistice Day, 11 November, the day the First World War officially ended. More brilliant still, she took the poppy as a symbol. It became acceptable and then important to wear an artificial poppy – made from cloth or paper, because poppies don't bloom in November – on Armistice Day, which became known also as Poppy Day.

The poppy remains a fundraising tool. It is owned – the phrase is legally accurate – by the Royal British Legion. This organisation supports current and former members of the armed forces and their families. The poppy campaign for 2018 raised £50 million. Though the remembrance poppy crops up in other countries, it is seen, certainly in Britain, as quintessentially British, even though the idea came from a Canadian and a Frenchwoman.

Poppy-wearing has grown more and more important. Poppies sprout from the lapels of politicians as early as mid-October; newsreaders on television wear them from the last week in October, closely followed by everyone who appears on television for any reason whatsoever. The period of mandatory poppy-wearing is getting longer. People who appear on television without a poppy will be criticised: in some newspapers vilified for their failure to show respect to the fallen. My grandfather fought in the Salonika Campaign in the First World War and reached the rank of sergeant. He would have been appalled by the idea of the compulsory poppy; he would certainly have refused to wear one.

The power of the poppy as a political symbol lies in its claim to be nothing of the kind. In 2011 the international football (soccer) organisation FIFA attempted to prevent the England team from wearing a poppy symbol on their team shirts for a match against Spain close to Armistice Day. The outrage in Britain was so colossal that FIFA backed down. David Cameron, when prime minister of Britain, was on a state visit to China at the poppy time of year in 2010. The Chinese asked him to remove his poppy, which has connotations with the Opium Wars of British imperialism (see again Chapter 57). Cameron refused: to do so was, perhaps literally, more than his job was worth.

The British newsreader Jon Snow courted controversy by refusing to wear a poppy and complained of 'poppy fascism'; he was routinely condemned in the expected places. Some Muslim groups refuse to wear poppies, seeing them as a symbol of British imperialism.

But for all that, the poppy has a power over the imagination, especially the British imagination. The year of 2014, the centenary of the start of the First World War, was marked by an installation at the Tower of London, in which the moat was filled with ceramic poppies: 888,246 of them – the number of soldiers from Britain and the British Empire killed during the war.

We take many powerful symbols from the natural world because the human mind responds to them so vigorously, filling them with meanings, many of them contradictory. Plants fill our minds: plants wanted and unwanted, plants cultivated and plants routinely killed when they get in our way, are all part of the furniture of our minds. Right and wrong, life and death: all summed up in a weed.

TWENTY

PAPYRUS

Reading maketh a full man; conference a ready man; and writing an exact man.

Francis Bacon

The poppy represents history. The papyrus plant *is* history. It's no symbol: history has been recorded on and made from its thick and sticky pith. The direct transfer of knowledge through time – unpolluted by the errors that come from transmission by word of mouth – was made possible by this species of sedge. For the first time, huge amounts of knowledge could be stored, accessed at will and passed on through generations.

Papyrus is so intimately associated with written knowledge and depictions of myth that we almost forget that it's also a plant that grows, photosynthesises, and supports life. The first time I looked on a wet, bright green papyrus swamp it seemed bizarre without its animal-headed gods and the stern profiles of emperors. It seemed the most unlikely source of a medium that would record and change history. The vivid green colour was set off by birds called red bishops: for them, papyrus was a breeding opportunity: the chance to pass on their genes and so seek immortality. The Ancient Egyptians recorded their own notions of immortality on fabric made from the same plant.

The annual flooding of the Nile created huge wetlands filled with papyrus. The plant grows tall: 16 feet (5 metres) high. Papyrus tends to dominate a landscape, or a waterscape. For the Ancient Egyptians papyrus was accessible in huge quantities. Naturally it was put to all kinds of use: boats, mats, rope, sandals and baskets. You could use it to feed domestic animals. The plants spring from rhizomes, which are modified stems that grow horizontally underground, putting out roots and new shoots as the plant spreads; these rhizomes can be eaten by humans.

Papyrus was crucial to the way the people of Ancient Egypt understood life and their place in the world. Papyrus plays a part in the creation myth, which is about the appearance of land from darkness and flood; this is an event that, before the building of the Aswan Dam, happened along the Nile on an annual basis. There is a myth in which Isis hides in the papyrus swamps with her son Horus, to save him from his brother Seth, who had already killed her husband Osiris and usurped the kingdom. A papyrus swamp is a landscape of mythologies, and also one of great

From swamp to library: papyrus in an Ancient Egyptian wall-painting from the tomb of Akhenaten, c.1375 BC.

practical importance. It was a plant that was central to the civilisation of Egypt, from which European civilisation can trace its origin. But papyrus is most significant as a medium for the transmission of history and the understanding of life and death. It was important because the Egyptians invented a system for recording such things. It is called writing.

The Egyptians may or may not have been the first to do so; best leave that question to the scholars; it's possible they got the idea from the archaic scripts of Sumer. It's also possible (perhaps even likely) that writing was invented by other civilisations independently. What's certain is that Egyptian hieroglyphics were up and running by the fourth millennium BC, and that you can find coherent texts dated from 2,600 BC. It's not just pictorial script; it's real writing: you need to know the language before you can understand the script.

Hieroglyphics were originally written on stone; the name means 'sacred carving'. A more portable medium for written information came with the development of suitable stuff derived from the papyrus plant. The earliest known papyrus with written text is the Diary of Merer, from 2560 BC, which some believe describes the building of the great pyramid of Giza. Papyrus is prepared for writing by removing the outer leaf of the plant to reveal the sticky and fibrous inner pith. This is then cut into strips; the strips are then set out in two layers and hammered together while wet, before being dried under pressure. The resulting fabric is polished with a round object: stone, shell, wood.

The process is both labour-intensive and skilful. The surface presented is irregular and difficult to work with. The scribes wrote with a piece of thick reed, and later a sharper implement called a stylus. Papyrus was for the elite, and the scribes were the curators of knowledge.

The recording of information changes everything. You are no longer reliant on the authority of individuals. You can be certain about who said what, and when they said it. You no longer rely on what people could remember: you can go to the source and look it up. Anyone who could read is, at least in theory, as wise as anyone who has ever lived. The Ebers Papyrus, a scroll 65 feet (20 metres) long and filled with medical text, was simultaneously practical and magical. You could also read *The Egyptian Book of the Dead*, which contains the most famous papyrus image of them all, that of the jackal-headed god Anubis weighing the heart of the scribe against a feather.

But no society is all about the great and the grandiose, not even Ancient Egypt, obsessed with both. There are surviving papyrus sheets with information about household matters and administration, also letters, contracts, legal texts, illustrated stories like *The Tale of the Shipwrecked Sailor*, and religious texts. All human knowledge was there: it could be written down in hieroglyphics on papyrus and consulted later for as long as the papyrus lasted: and in Egypt's mostly dry climate the chances for preservation are good.

Hieroglyphics were mysteries and curiosities until 1822, and the deciphering of the Rosetta Stone. This contained the same information – a decree of King Ptolemy V – in three different scripts: hieroglyphics, demotic Ancient Egyptian and Ancient Greek. Scholars were now able to read what the people of Ancient Egypt wanted to record.

The sedge plant of papyrus allowed one of the great civilisations to develop, and to spread its influence across the ancient world. The civilisations of Greece and Rome have their debts to Ancient Egypt. The Egyptians created an abiding literary culture: and the civilisations that followed adopted and adapted that model. The existence of a tall and near ubiquitous sedge has a great deal to do with that process.

TWENTY-ONE

PENICILLIUM

Nature makes penicillin. I just found it.

Alexander Fleming

How many readers of this page would be dead without the genus of fungus known as *Penicillium*? How many people in the world would be dead? How many people would never have come into the world without this fungus, their parents only able to survive and breed because of the actions of that fungus? It's possible, perhaps even probable, that this book would not have been written without it, its author incapacitated or worse. The world would be a very different place without the fungus in it: one with far fewer people, in particular far fewer children and far fewer old people.

There are more than 300 known species in the genus *Penicillium*, and they are present in the air and in dust. They can spoil food. They are useful in cheese-making: essential to Camembert, Brie, Roquefort and a good few others. Some species can damage machinery, oil, fuel and protective and optical glass.

And some can cure illness in humans by attacking pathogenic bacteria. Across the millennia, pneumonia and diarrhoea were two of the biggest killers of humans. All at once they became curable; we now think of such illnesses as inconveniences rather than life-threatening conditions. The steep rise in the human population of the earth after the Second World War coincided with the Golden Age of Antibiotics. Life on earth is unrecognisable from what came before: in 1919, the pandemic known as Spanish Flu killed 50 million people.

'One sometimes finds what one is not looking for,' Alexander Fleming wrote some years later. 'When I awoke just after dawn on September 28, I certainly didn't plan to revolutionise all medicine by discovering the world's first antibiotic, or bacteria killer. But I suppose that was exactly what I did.'

Fleming was researching bacteria, and how to kill species harmful to humanity. He was doing so at St Mary's Hospital in London. The much-told story of the accident that led to his discovery of penicillin is pretty well true. In 1928 he returned to work after a month's holiday in Suffolk. When he reached his basement lab he found that he had failed to cover a petri dish containing *Staphylococcus aureus*, a species of bacteria that can affect the skin and mucous membranes of mammals, including humans, and birds. And it had gone mouldy, as things do – or,

Author of the breakthrough: Alexander Fleming (1881-1955), discoverer of penicillin.

to put that more accurately, a fungal growth had established itself on the dish. Fleming noticed that the bacteria had died back around the new growth, and explored the further possibilities of this mould.

Fleming wasn't the first person to notice that mould could help humans to combat infection. In Ancient Egypt (as you can read in the previous chapter, on papyrus) they applied mouldy bread to wounds to help them heal. In the nineteenth century scientists had worked with mould as a healing substance, but without starting any medical revolutions. Among the problems, as Fleming himself found, is that the right kinds of mould are hard to grow, and it's very difficult to isolate the active parts of the mould and keep them stable.

Fleming's discovery was not an overnight business. He first thought he had found a useful disinfectant. Subsequent work showed that the isolated ingredient, which he called penicillin, was not toxic to humans. But there were problems with producing the stuff in quantity, and the revolution was put on hold. It required two other scientists to make Fleming's discovery effective. The picturesque story of the chance discovery of the medicine that changed the world has given great pleasure to millions, but it would have come to nothing without Howard Florey and Ernst Chain at the University of Oxford. By 1940 they were curing mice, and a year later penicillin had worked successfully on humans. The three of them shared the Nobel Prize for medicine in 1945, amid resentment about the prominence given to Fleming.

Florey and Chain took their work to the United States in 1942. Mass production was initially very difficult, but this goal was achieved in time to treat American troops at the invasion of Normandy in 1944. The most effective strain for the manufacture of penicillin was found on a cantaloupe melon.

Australia became the first country to make penicillin available to civilians, and so the golden age began. It is memorably described in the books by James Herriot, the fictionalised memoirs of a vet in the Yorkshire Dales whose real name was James Alfred Wight. He describes his first experience with antibiotics: 'like witnessing a miracle'. It really did seem then that humans had cracked it: that illness was now history.

Penicillin worked on the bacteria that caused scarlet fever, pneumonia, meningitis and diphtheria. New and still more effective antibiotics were discovered. Antibiotics were increasingly used for treating farm animals, not just as treatment for ailments but as preventative measures, to promote growth and maintain herd health. The animals could now be farmed in much closer proximity. It was the beginning of factory farming. It gave the world cheap meat, which is now considered a human right in many nations; it's been estimated that 73 per cent of the world's antimicrobial medication is used for farm animals.

The golden age couldn't last. The reasons for this can be confusing. Some people believe that humans become immune to antibiotics from overuse. Others believe that the bacteria acquire immunity by subtle and devious mutations. The issue is simpler. The golden age of antibiotics was a stunning exercise in the killing of bacteria, but some survived. There remained sub-populations of bacteria immune to certain antibiotics. Just as farmers selectively breed cattle for their meaty yield, so we were – quite inadvertently – selectively breeding bacteria for their immunity to penicillin. And we have continued to do so. Immune strains of pathogenic bacteria are increasingly common. (I should point out here that

Covid-19 is a virus, not a bacterium, which means that it can't be treated with antibiotics. Bacteria are free-living organisms that can survive inside or outside a living body; viruses are groups of molecules that can't survive without a host – as Paul Nurse says in his book *What is Life*, viruses 'cycle between being alive, when chemically active and reproducing in host cells, and not being alive, when existing as chemically inert viruses outside a cell'.)

A report from the World Health Organization in 2014 suggested that we are facing a crisis. 'Without urgent, coordinated action by many stakeholders, the world is headed for a post-antibiotic era, in which common infections and minor injuries which have been treatable for decades can once again kill,' said Dr Keiji Fukuda, WHO's Assistant Director-General for Health Security. 'Effective antibiotics have been one of the pillars allowing us to live longer, live healthier, and benefit from modern medicine. Unless we take significant actions to improve efforts to prevent infections and also change how we produce, prescribe and use antibiotics, the world will lose more and more of these global public health goods and the implications will be devastating.'

The use of antibiotics as prophylactics rather than cures for farm animals was set to be banned in the European Union from 2022 as awareness of the reality of the impending crisis has spread. But this practice is essential in many forms of intensive farming, especially in the United States. With the centenary of the discovery of penicillin in sight, we are looking at a future without antibiotics. The post-antibiotic age lies ahead of us. The loss of such medication will change the world as drastically as their discovery. A fungus made modern life possible; the loss of its effectiveness will be the start of another era. And it has already started.

TWENTY-TWO
KIGELIA

Bring me my bow of burning gold:
Bring me my chariot of fire.

William Blake

A decent-sized kigelia tree looks like an Italian delicatessen: dominated by hanging sausages of improbable size. The sausages are fruit; they can weigh up to 10 kilos (22lbs) and the plant is colloquially known as a sausage tree. Don't dally beneath it: plenty of people have been injured and even killed by these

botanical bombs. The trees grow across the African savannahs, mostly along watercourses. The flowers and fruit are an important food for many of the creatures that live there, and the tree has a long history of uses for humans. And while all this makes the kigelia an interesting tree, it's not why it has made it into these pages. The kigelia is here because it allowed humans to take over the planet.

Here is the toolkit that humans used for global conquest: one flat piece of kigelia wood, one long – say 2 feet (60cm) – twig from a potato bush, and a handful of dried elephant dung. Potato bush is a savannah plant with small flowers that emit scent in early evening to attract moths as pollinators. The smell is just like baked potatoes, though the plants are not closely related. The potato bush is useful because you can take from it a slim length of very hard wood. You pair this with a small flat slab from a kigelia. Other kinds of wood will do the job, but kigelia is the best: it has just the right flaking softness for the job.

Fire-bringer: Under the Sausage Tree *by Shelly Perkins, showing a kigelia tree in South Luangwa, Zambia.*

Finally you need your elephant dung. This is not even remotely disgusting; elephants digest plant matter in a rough and ready fashion: a dry elephant turd is just flaked and minced vegetation, with no odour to speak of. It's easy to find in the dry season; in the wet it's advisable to keep a personal supply of dry dung safe from the rain. Spread a little of this stuff; it looks like old tobacco. You are now ready to begin.

Rest your potato bush spindle on the kigelia platform and spin the spindle back and forth between the fingers and the palms of your two hands, which you present to the spindle as if in prayer. With use your platform acquires a series of small holes, worn by your spindle. Normally you reuse one, which is quicker than starting a new one. As you spin your spindle, repeatedly work your hands down its length, pressing downwards: the downward pressure of the spinning hardwood onto the resting softwood is what makes it all happen. In much less time that you would believe possible, you see smoke rising from the kigelia platform; you smell it too. The action of the spindle on the bottom of the hole has caused small fragments of kigelia to break off into tiny hot coals. You tip these onto your elephant dung and blow – and then, as if by the intervention of God, you have flame. You have fire. You have power. You have control. You have the whole earth before you.

There is no consensus as to when our ancestors made and controlled fire. That's partly because it was a gradual process, beginning with wildfires, and a growing ability to exploit them. The bounty of wildfire, caused by electrical storms, trapped animals in the flames, animals that could be gathered and consumed, giving humans a taste for cooked food, which is easier to chew and to digest. There is a rough and ready marker of 1 million years ago for control of fire: and it's likely that kigelia and potato bush played a part in this – but in a way the species don't matter. It's all about mastering the principle of making fire from plants.

Fire changes everything, of course. The light made nights less fearful, and besides, fear of the fire kept the frightening animals away. If human enemies came, at least you could see them: the unknown is always more frightening than the known. The heat from fires increased every individual's chances of survival, and it also meant that humans could now survive in colder places. The earth was opening up to the species that could control the flames.

Cooking made a great many more food items available to humans, because cooking made them digestible. It's been speculated that the increased availability of food by means of fire fed the growth of the human brain. Humans learnt how to make fire because they were smart, and once they had fire, it made them smarter: a pleasing idea. Fire was (and is) an effective way of managing land: stimulating new growth that attracts grazing animals. Fire also made it possible to make better wooden tools, for they could now be hardened in the flames. What's more, meat could be dried over a fire and kept for a good deal longer: opening the future to those who could make plans.

So from the action of one plant against another, and with the help of plants already consumed (by elephants), humans were able to take the biggest step in their development so far, perhaps the biggest ever. Fire allowed humans to make darkness a productive time: times that could be used for communicating, socialising, singing, praying: stuff beyond the simple need for survival. What humans first did at noon in the shade of a strangler fig they could now also do at night, by the light of the fire.

Humans could pack more into a day than an animal without fire: sixteen daily hours of wakefulness is considered normal for humans – far more than most other species. Our closest relations, chimpanzees, rise with the dawn and retire at dusk: a twelve-hour day in the tropics. Fire gave humans the chance to do more things and to do them socially. There have been two great advances in human culture: the second is the invention of agriculture; the first is the control of fire. Besides these, all other steps are incidental – save the twin human talent for both construction and destruction.

We are who we are because of the kigelia. It's a fine tree in many other ways, and it's found all over tropical Africa – also to an extent in India, where it was spread by humans. The preposterous fruit are much eaten by hippos and rhinos where they still exist, and the seeds are spread in their dung. The trees aren't the tallest, reaching around 66 feet (20 metres) maximum. They have wide, generous spreading branches, much favoured by resting leopards, and their shade is almost as inviting as the strangler fig. Avoid that invitation; I know someone who was unconscious for three days after driving beneath a kigelia; laid out by a falling fruit. The flowers are huge red trumpets that form essential dry season nutrition for herbivores in many places; baboons and impalas routinely gorge on them.

In the Okavango Delta they make *makoros*, punted canoes, from kigelia (but see Chapter 42 on the marula tree); the Kikuyu make beer from the fruit. Material from the tree has been used to treat wounds, ulcers, syphilitic sores and skin complaints like psoriasis and eczema. The plant has also been used to treat skin cancer; a friend of mine acquired considerable relief from cream made from kigelia. Kigelia is also used in cosmetics and so-called anti-ageing creams. The tree is a curiosity for visitors to Africa, who make their safari and marvel at the tree with the thrillingly dangerous fruit. But it's also a hidden part of our history, its importance all but forgotten – until people who have retained the skill choose to demonstrate the ancient art of fire-making, the hot little kigelia coals are poured onto the waiting dung, there to receive the encouraging breath: and once again there is fire.

TWENTY-THREE
DAFFODIL

And then my heart with pleasure fills,
And dances with the daffodils.

William Wordsworth

The beauties of the wild world are obvious to us all. Now. We take it as self-evident that a rainforest is beautiful, that a river lined with willows is beautiful, that a bluebell wood is beautiful. But this idea of natural beauty is a fairly recent development and in some societies it has hardly developed at all.

Nature has been, at least since the invention of agriculture, as much an opponent as a lover. There was so much of it, it was overwhelming; it was perpetually threatening; people took it for granted and saved their reverence for the works of humankind. Later on, as we became city-dwellers, we developed a taste for gardens, for a bit of nature in the middle of town. Nature was now fine – so long as it was 'in its proper place'. That is to say, quite obviously under human control. Natural beauty was acceptable so long as humans dictated what grew and what did not.

Lionel Brown – Capability Brown – created natural-looking landscapes designed to please human eyes: to give people a feeling of peace and control, of both belonging and owning: a grazed sward, mature trees, open views with access to water, a place where a person can feel safe, full with confidence about the future. Alexander Pope wrote:

...let nature never be forgot,
But treat that goddess like a modest fair,
Not overdress, not leave her wholly bare.

Letting nature go its own way was anathema. But with the advance of the industrial revolution, more and more nature was destroyed, and more and more people lived in cities. Meanwhile, better communications – roads and increasingly railways – allowed people in cities to reach wilder places more easily than before. With destruction came revelation. Nature without the hand of humanity had its meaning, and perhaps the greatest meaning of all... which brings us to daffodils and William Wordsworth's ecstatic celebration of wild nature. His daffodil poem was first published in 1807, though the experience that is recorded took place in April 1802.

Inspiration: The Young Poet, *1849, by Arthur Hughes (1832-1915).*

Daffodils have their origin on the Iberian Peninsula. They belong to the genus *Narcissus*, which is always associated with the myth of Narcissus, the beautiful youth who fell in love with his own reflection and killed himself. 'Self-love so often seems unrequited,' a character in Anthony Powell's *A Dance to the Music of Time* observes.

The plant has also been used in medicine. It induces vomiting, and numbness, and has been used as a purgative and a treatment for burns and wounds. It can cause hallucinations; Sophocles called the plant 'the chaplet of the infernal gods'. Hippocrates recommended it as a pessary for uterine tumours. A substance called galantamine has been extracted from the plant and is used to treat early stages of Alzheimer's dementia.

The dramatic good looks of daffodils have always been important to humans, perhaps especially because of their forwardness: the way they are, in many places, the first large flower of the year to show itself. Daffodils are also ridiculously easy to cultivate: all you need is a bulb. Botanically, this is a modified stem that is used to store food during dormancy, giving the plant its lightning-swift response to the onset of spring. The plant propagates vegetatively (clones itself) by division of the bulb, and it also does so by seed, relying on insect pollination.

The plant has been spread and introduced by humans across the centuries. There are around 26,000 cultivars from seventy-four species of narcissi; many of these grow feral, sometimes in the fantastic varieties that some people love to cultivate. But the species sometimes called Lent lily, with the confusing scientific name of *Narcissus pseudonarcissus*, has a good claim to be native English. These, most people agree, are the ones Wordsworth wrote about. In Shakespeare's play *The Winter's Tale*, Perdita speaks of

> *... daffodils*
> *That come before the swallow dares and take*
> *The winds of March with beauty...*

Which prefigures Wordsworth's daffs very sweetly. We always associate the romantic experience, as captured by the Romantic poets, as one of perfect authenticity, and Wordsworth's poem begins, as everybody knows, 'I wandered lonely as a cloud'. But he wasn't lonely: he was with his sister Dorothy – and she was the one who first wrote about daffodils. They found them in the course of a wild walk around Glencoyne Bay in Ullswater in the Lake District. The walk took place in 1802, and Dorothy wrote in her journal: 'I never saw daffodils so beautiful they grew among the mossy stones about and about them, some rested their heads upon these stones as on a pillow for weariness and the rest tossed and reeled and danced and seemed as if they verily laughed with the wind that blew upon them over the Lake, they looked so gay ever glancing ever changing.'

Her brother's poem was published five years later in Wordsworth's *Poems, In Two Volumes*, and here too the daffodils danced, and if they didn't laugh out loud they were 'jocund company'. The critics didn't think much of the collection; Lord Byron reviewed it and gave it a stinker, but after a slow start it caught on with the mere public. The daffodil poem became a staple of anthologies and the school curriculum. It has given us the romantic idea of Romanticism: the poet wandering fecklessly about the countryside, retuning to shelter only to throw out a poem in a spasm of genius. The fact is that this poem, like most other good poems, is the result of hard work, revision and a little artful artificiality.

The poem has become a classic partly because it is a beautifully made poem, and partly because it caught the wave: the revolutionary idea that true beauty lies in the natural world, away from the works of humankind. Coming upon a new vista and being socked between the eyes by pure beauty – well, it's an experience known to us all. No poem has ever captured it better. And this was wild beauty: with the unspoken corollary that the work of humans is ugly.

The more scarce a thing becomes the more we are inclined to cherish it. John Keats mentioned daffodils – perhaps a courteous nod to the older poet – in a list of delightful things including 'daffodils/With the green world they live in'. That's from *Endymion: A Poetic Romance*, which begins 'A thing of beauty is a joy for ever'.

That last line has gone so deeply into the language that it is impossible to hear it for the first time; the same is true of Wordsworth's opening line. A 1980s television advert began: 'I walked about a bit on my own… oh no…' before a can of beer is consumed and the narrator is able to write the line that everyone knows. ('Heineken refreshes the poets that other beers cannot reach.')

The poem about the daffodils, and with it, all living daffodils, has become an instantly recognisable idea – about mad romantic poets, yes, but also about love. Love of landscape: love of wild landscapes full of wild plants. It is a love fuelled by the increasing scarcity of such landscapes.

Ten thousand saw I at a glance
Tossing their heads in sprightly dance…

TWENTY-FOUR
APPLE

The serpent beguiled me, and I did eat.

Genesis 3:13

The apple is the default fruit of Western culture. When we hear the word fruit, we assume that it's an apple until or unless told otherwise. It was assumed that the forbidden fruit of the Bible (and the Qur'an) was an apple, though neither text is clear about the species consumed: other possibilities include grape, pomegranate, fig, carob, citron, pear and even a mushroom. The apple has also become a beloved symbol of the United States and a nickname for its greatest city. America likes to see itself as a nation built on pioneering spirit and apple pie. Other uses to which the pioneers put apples will be examined shortly: what you might call the true pioneering spirit.

Apples were one of the first fruits to be domesticated, not least because even wild apple trees bear large fruit. (There are more than thirty species of wild apple, only one of domesticated apple.) Note that it's much easier to domesticate plants than animals. Domestication is a difficult process with all animal species, and frequently dangerous with large mammals. But plants don't move, they don't need food, they don't bite and they don't trample people underfoot. They can be domesticated by anyone who can dig a hole in the ground, or, for that matter, anyone who can claim a good tree for his own. That's why there are thousands of species of domestic plants, but only a few dozen of domestic animals.

No one can say for sure when apples were first domesticated: perhaps as long as 10,000 years ago. They came from Central Asia, travelling to Europe along the Silk Road. But there's a problem with apples, as with many domesticated plants: they don't breed true from seeds. Or, to put that another way, a seed from an apple may not grow up exactly like the tree it fell from. A seed from a tree bearing large, sweet fruit may produce trees with small, sour fruit. It's an example of life's natural tendency to diversity; the human tendency has always been the creation of monocultures.

The solution to the problem is grafting: putting the tissues of two different plants into contact. You attach twigs from one plant – genetic material, we would

Opposite Knowledge of good and evil: Eve, 1896, by Lucien Lévy-Dhurmer (1865-1953).

say these days – onto a growing tree, the rootstock. And it grows readily enough in the right conditions and produces exactly the fruit of the tree that the grafted material came from. Thus an apple variety can be propagated truly, apple trees are comfortable with this technique and humans are delighted by apples: it's been estimated that there are now 7,500 apple cultivars in existence.

Apples have another advantage: they keep pretty well. If you can find a good storehouse a little above freezing, you can keep apples all winter. Fresh food in winter was a very important thing, nutritionally and psychologically. Apples, with good quantities of vitamin C and a refreshing taste, kept generations of humans going through the cold months. In days before refrigeration and airfreight, apples brought a little summer into winter.

Apples were fruit that mattered. For one of his labours, Hercules had to pick apples from the tree of life in the Garden of the Hesperides. Paris was required to present an apple from the same garden to the goddess he found most beautiful, a procedure that was always going to lead to trouble. Paris was finding the decision difficult, so the goddesses obligingly stripped naked to help him make up his mind. Hera, Athena and Aphrodite all offered bribes; Paris gave the apple to Aphrodite after she offered him the most beautiful woman on earth, Helen of Sparta. He took her off to his home town where she became Helen of Troy; the Greek army and ten years of war followed.

The most famous apple story is that of the Garden of Eden, in which first Eve and then Adam succumb to the temptation to eat from the tree of knowledge of good and evil. St Jerome created the Vulgate, the Latin version of the Bible, in the fifth century AD and it's possible that the apple tradition came from this: the tree gave knowledge *boni et mali*, of good and evil. *Malus* is also Latin for apple – though perhaps apple was always assumed to be the fruit in question. The naked woman offering an apple to a naked man with a snake in the background: it's one of the archetypal images of Western civilisation.

Apples became important commercially. Inevitably they went with the pioneers to North America. One man gets most of the credit for spreading apples around the continent: John Chapman, who lived 1774-1845 and became a semi-legendary figure known as Johnny Appleseed. The most picturesque versions of his story have him walking in a perpetual pilgrimage around the countryside with pockets full of seeds that he planted wherever he went, creating vast orchards and bringing great good to the emerging nation. He cured a lame wolf and the wolf accompanied him everywhere. He put out his fire because it harmed a mosquito. He travelled from household to household, sleeping on the floor and telling unforgettable stories.

In truth Chapman established nurseries rather than planting random orchards. He was a proselytiser for Emanuel Swedenborg's version of Christianity. He was certainly a remarkable man, and he brought a lot of apples into the American countryside, consequently a lot of apple pie. But there is a secondary (or perhaps a primary) use for apples: they turn into booze. You can hardly stop apples turning into cider (as an alcoholic drink rather than in the American sense of fizzy apple juice).

In the hard winters of the northern parts of North America, liquids left outside will freeze. If you leave your cider outside and regularly skim the ice from its surface, you will slowly distil the cider and create – applejack. That's a spirit with a very acceptable kick, and it's as helpful in getting a person through the winter as any sweet fruit. The process of distillation by freezing doesn't get rid of the dangerous methanol; the resulting ferocious hangovers are known as apple palsy. Applejack was eventually outcompeted by commercial (steam-distilled) whiskey, but the pioneers of early America were helped through the dark times by applejack.

Myths accumulate round apples. Sir Isaac Newton was supposedly sitting under a tree when he was hit on the head by a falling apple – and instantly grasped that the force that drove the apple to the ground also kept him safe on that ground, the moon from falling into the earth and the earth from falling into the sun. The story is mostly true, barring the blow on the head: one of his first biographers, William Stukeley, wrote: 'He [Newton] was just in the same situation, as when formerly, the notion of gravitation came into his mind. It was occasion'd by the fall of an apple, as he sat in contemplative mood.'

Apples seem to us both homely and mysterious, everyday objects with aspects beyond easy understanding. Apple was the name given to the Beatles' company, founded in 1968 under the punning title Apple Corps, of which Apple Records was a subdivision. It was initially an experiment in idealistic capitalism; I knew someone who claimed to have been employed by Apple in the post of poet. In 1976, Steve Jobs, Steve Wozniak and Ronald Wayne founded Apple Computer Company, now Apple Inc., and as a result had a long series of legal disputes with Apple Corps. The homely/mysterious nature of the name and the fruit helped their computers to gain acceptability.

Sin, eternal guilt, perpetual discord, winter salvation, a vast nation's self-image, the point at which the wave of the 1960s counter-culture broke and rolled back, and all that is slickest about hyper-modernity: that's apples for you. It's the fruit that means all things.

TWENTY-FIVE

REDWOOD

From the redwood forest to the Gulf Stream waters
This land was made for you and me

Woody Guthrie

When we start giving names to individual trees, it's clear that there's something odd going on. We give names to individuals of different species of our fellow animals all the time: pets, obviously, but also animals that are the subjects of long-term study by ethologists, like Jane Goodall and the chimpanzee David Greybeard. The global outrage over the lion slain in 2015 by the trophy-hunting dentist was occasioned by the fact that the lion had a name: Cecil.

The urge to bestow names on trees is found most often with the redwood trees of the United States, huge conifers that belong to the subfamily *Sequoioideae*. There are trees called Hyperion, Helios, Icarus, Daedalus, Wawaona Tree, Pioneer Cabin Tree, Chandelier Tree – and, perhaps most tellingly of all, General Sherman.

Redwoods are huge. Individuals above 300 feet (90 metres) are not remotely unusual, and a diameter of around 20 feet (6 metres) is standard. They grow in northern California and southwest Oregon and hold most of the records for the tallest and the heaviest living trees; examples and numbers to follow. The cones they bear are modest things little more than an inch (2.5cm) in length. They are able to grow tall because the local conditions favour such extravagance – and, as always, the tallest trees have enhanced access to life-giving light. There are always advantages for a tree that can outgrow the rest while remaining stable.

They exploit many helpful factors: the area has a good deal of rain: 60 to as much as 140 inches (150-350cm) in a year. The places where they grow are notoriously susceptible to summer fogs and the damp, dense atmosphere means that the trees lose very little liquid from transpiration and photosynthesis. The rich alluvial soil, packed with nutrients, helps to fuel the growth of these super-trees. The combination of the strength of these trees and the moist atmosphere has made them a rich habitat for epiphytes; that is to say, plants that grow on other plants without otherwise taking advantage of them. Living trees found on redwoods include a Sitka spruce 40 feet (12 metres) tall.

There are three genera in the subfamily of *Sequoioideae*; sequoia is also the name of a single species within that group. The name was chosen by the polymathic

Stephan Endlicher, a nineteenth-century Austrian philologist and botanist. He chose the name to honour Sequoyah, who invented the Cherokee writing system. It was a way of expressing admiration both for the individual and for the trees themselves: such botanical immensity demands respect.

They are trees of an ancient lineage that goes back to the Jurassic era. Ancestral redwoods have been found in Europe and China as well as in North America. In 1848 there were about 2 million acres (81,000 hectares) of coastal redwoods. That changed with great rapidity the following year: the year of the gold rush, commemorated in the song about my darling Clementine and in the name of the San Francisco NFL team, the 49ers.

With this sudden and dramatic population shift, the trees were now seen as a glorious resource, and many were felled for building, and a new West was created from the ruins of the old. Around 5 per cent of those nineteenth-century forests remain: 100,000 acres (40,000 hectares), spread in a series of patches along the coast, protected in thirty-one state and national parks.

It is clear, then, that there has been a certain amount of revisionism with redwoods, representing the change in thinking that spread across much of the world in the course of the past century or so. The easy acceptance of nature's apparently infinite bounty has been replaced by an urgent understanding of the fragile and finite nature of natural resources.

Perhaps the most poignant shift in understanding has come with the famous drive-through redwoods. I remember being enchanted by pictures of these trees when I was a boy. The idea that you could drive a car, a properly massive American truck, right through the middle of a tree: that seemed to demonstrate both the glory of the natural world and the American propensity for hugeness. On one side of the country you could look down from the Empire State Building; on the other, you could drive through a tree and look up at its impossibly distant branches: what a country.

I imagined that these trees became tunnels out of necessity: that you had to tunnel through them if you wanted to get anywhere. It came as a disappointment to realise that these drive-through trees were tunnelled out for wholly frivolous reasons: to amuse human beings, to impress them with nature's might and, at the same time, to put nature's might in its place. Trees, like everything else, were there to amuse people, or what was their point? And so a number of trees were laboriously pierced, and a million pictures were taken of the vehicles passing through.

The Wawona Tree in Yosemite National Park was perhaps the greatest example. It stood at 227 feet (69 metres) and was 25 feet (7.5 metres) in diameter at the base. It was dug out in 1861 by the Yosemite and Turnpike Company as a tourist attraction: the idea being to bring more people to the National Park. That was perhaps a good and important ambition; certainly the network of national and

state parks in America is a great thing that attracts many visitors, so in terms of conservation and human pleasure, the establishment and maintenance of these parks has been a great success. But the Wawona Tree was a casualty: fatally weakened by the tunnelling, it fell after a storm in 1969: more than 2,000 years of life ended by human silliness. Other tunnelled trees survive: you can savour the Chandelier Tree, 276 feet tall (84 metres), in the evocatively named Drive-Thru Tree Park: and no, should I ever go there I doubt if I would be too high-minded to stop and be photographed, if only in homage to my childhood. I'd probably say it was ironic.

Revisionism is part of the human condition: each generation sees the world differently and seeks to make such shifts in morality retroactive. This book will come up against such tendencies again and again: so let us briefly pause at the tree known as General Sherman. It stands in the Giant Forest of Sequoia National Park, and is claimed to be the biggest tree on earth, if not quite the tallest or the widest; it has a bole volume of 52,513 cubic feet (1,487 cubic metres). It is between 2,300 and 2,700 years old. It is named for the great American hero, the Civil War general who captured Atlanta and then directed his nation through the triumphant Indian Wars – or, if you prefer, the brutal perpetrator of scorched-earth policy and the violent suppressor of American indigenous people. It all depends on your revisionism.

These days, the trees get names that are less likely to be politicised or otherwise revised. The tallest of them all is Hyperion, named for the Greek mythological figure who was one of the twelve Titans. The name means 'high one': not inappropriate for a tree that stands at 380.1 feet (116 metres). The locations of Hyperion and other giants are kept secret, to protect them from vandalism and because, as a website I consulted in the course of researching this chapter added gratuitously, 'people are jerks'.

Opposite Big Tree Wawona: *in Yosemite National Park, USA, pictured c.1880; it stood till 1969.*

TWENTY-SIX
CHRISTMAS TREE

Oh Christmas tree, oh Christmas tree
Thy leaves are so unchanging.

Anon

Few of us who live in the richer countries of the northern hemisphere think of ourselves as tree-worshippers. And yet most of us take a tree into our homes at the winter solstice every year, decorate it, make much of it and use it as the centre of celebrations that last for two weeks and more. We take trees into our homes at the very darkest time of year: by the time we take it out again, the days are getting longer.

As just about every page in this book will tell you, sun plus water equals life. But in the drastically seasonal lands of the northern part of the northern hemisphere, we take water for granted. It's the sun that comes and goes, so the sun seems the great giver of life. The early humans who sat beneath the shade of the strangler fig (see Chapter 1) to escape the midday sun of tropical Africa had another view; in the hottest time of the year they longed for the return of the life-giving rains. But in temperate lands the increasing warmth of the sun triggers the growing period of the spring, nourishes the harvest and ripens the fruit; the shorter and colder days, the days with little sun, are for hanging on and waiting – waiting for the return of the sun, life and hope.

It follows that the December solstice has been a big deal ever since humans learned how to predict it. It is the moment at which the North Pole starts to tip back towards the sun; in England the daily allocation of eight hours of daylight at the winter solstice will, by the time of the summer solstice six months later, have swelled to more than sixteen. The thought that the worst of the darkness is over gives new strength in the hard times of winter.

And so people celebrate. One obvious way of doing so is to take living green boughs – even whole trees – into your home, as if you were bringing the life back in. There is evidence of such celebrations in Ancient Egypt, China, the Viking countries and among Hebrew people. The Ancient Romans celebrated the solstice with the Saturnalia, a time of ritual licence, feasting and lawlessness – and this often involved wreaths and garlands of evergreen plants.

The feast of the birth of Christ got tangled up with more ancient traditions of the December solstice. Scholars speculate about the real year of Christ's birth (between 2 and 4 BC is widely favoured) and the time of year. Some say that the shepherds would not have been abiding in the fields at night in December, because it would have been far too cold: the sheep would be penned and the shepherds in bed. It's also suggested that a census in which all the world should be taxed would not be held in winter, when daylight was short and travel problematic.

No one is quite sure how it happened, but the ancient celebration of the solstice and the celebration of the birth of Christ are now conflated: and the ceremony is marked by the veneration of evergreen trees. The process that made the tree a Christian symbol is not entirely clear. The Mystery Plays, traditionally performed on Christmas Eve, begin with the tree of paradise and the retelling of the story of Adam and Eve. Christmas trees were decorated with apples and wafers: the apples for the fall of humankind (see Chapter 24) and wafers for redemption through the Eucharist (see Chapter 2). There is a story that Martin Luther, delighted by the sight of a conifer with the stars behind it, cut a bough from the tree and took it home with him. Certainly, the Christmas tree began as a largely Protestant tradition, and centred in what is now Germany and the Baltic States. The Vatican didn't have an official Christmas tree until 1982, and Oliver Cromwell banned Christmas trees in Britain, along with carols.

Christmas trees were brought into royal courts in the nineteenth century. Charlotte of Mecklenburg-Strelitz, wife of George III, gave a children's party with a Christmas tree in 1800. Hans Christian Andersen's story 'The Fir Tree' was published in 1844. It's an odd tale; the little tree is cut down and decorated for Christmas, but is later left to moulder in an attic and then burned. But there is no doubt that the breakthrough date for Christmas trees was 1848. That year the *Illustrated London News* printed an engraving of Queen Victoria, Albert, the Prince Consort, and their children all celebrating Christmas around a decorated tree. Albert had brought the custom from his home in Germany and, following the royal lead, the British took it on with great enthusiasm.

Christmas trees were slower to spread across the United States, where imitating Queen Victoria was not high on the agenda. Franklin Pierce was the first president to have a Christmas tree in the White House in 1850, and the ceremony of the lighting of the White House tree was inaugurated in 1923.

What species makes the best Christmas tree? In the United States, Douglas fir, Fraser fir and Scots pine are the most usual; in Europe, Norway Spruce and Nordmann fir. Where do you get your tree from? In earlier times you went to the forest and cut one down, or if you lived in a city, you could buy one from a street vendor; it was an informal and opportunistic trade. The first Christmas tree farm in the United States was established in 1901 and demand accelerated. These days 35 million Christmas trees are sold annually in the United States, 300 million worldwide.

In 1965, the Peanuts cartoon *A Charlie Brown Christmas* was watched by 36 million people; it took the then-popular aluminium Christmas tree as a symbol of the commercialisation of Christmas. Aluminium trees became unthinkable at a stroke. But the popularity of plastic trees continues to, as it were, grow: a poll in 2004 showed that 58 per cent of households in the United States used an artificial tree.

Others reject this wholeheartedly: a real evergreen plant is needed for an authentic celebration of the winter solstice. The length of time a Christmas tree is permitted to stay in a house is controversial. In some traditions it mustn't be brought into the house until Christmas Eve, and mustn't be relinquished until Twelfth Night, 6 January. Other traditions require a tree at the beginning of Advent, four weeks before Christmas. In North America it's widely accepted that Christmas begins after Thanksgiving at the end of November, so that's when the tree goes up. As Christmas has become a festival of anticipation in Britain, Christmas trees tend to be introduced into a house in early December and thrown out around New Year. In some areas of London it is accepted practice to throw your tree into the street outside your house to await collection: pedestrians must pick their way on a narrow path between parked cars and abandoned trees.

Opposite *Tree of life:* On Christmas Day, c.1910, by Emile Czech (1862-1929).

TWENTY-SEVEN
VENUS FLY-TRAP

*'You don't seriously suggest that they're talking when they make
that rattling noise.'*

John Wyndham, *The Day of the Triffids*

Charles Darwin said the Venus fly-trap was one of the most wonderful plants in the world and he fed them on roast beef and hard-boiled eggs just to see how they coped. The plant has haunted the human imagination ever since it was widely known: and we have taken its carnivorous nature much further in tales of man-eating plants. Any vaguely threatening or intimidating plant is called a triffid, though comparatively few have read John Wyndham's 1951 novel *The Day of the Triffids*, or seen the 1962 film of the same title.

The Venus fly-trap has a relatively tiny global distribution; it's found only in swamps and wet savannahs in North and South Carolina in the United States. Europeans were aware of it by the mid-eighteenth century; the naturalist John Ellis wrote about it in the *London Magazine* and communicated with Linnaeus on the subject. Linnaeus was troubled: the plant didn't fit in with his view of the world.

His problem goes back to the ancient idea of the *scala naturae*. This was set out by Aristotle, but it's much older – and it's still prevalent. It's no longer a formal scientific concept, but it's very much a part of our informal notions of the way the world works. It's the idea that nature is hierarchical: at the bottom are rocks and minerals, which have no soul. Above them are plants, which have a vegetative soul but not a sensitive soul. (To this day we say that a person in prolonged coma is 'in a vegetative state'.) Above plants, separated by their ability to move, come non-human animals, some higher, some lower – you can still read the term 'higher animals' in scientific works – and over them, separated by their capacity for rational thought, humans. Beyond humans you find angels and archangels and at the top there is God.

That's why Linnaeus couldn't accept the idea that the Venus fly-trap ate insects, and believed that it sheltered them from the rain. The eighteenth century was the Age of Enlightenment, the time in which religion and science parted ways, but the process wasn't swift or easy or complete. Almost a century later Darwin delayed the publication of what became *On the Origin of Species* for more than twenty years because of the problems his ideas would make for religious people.

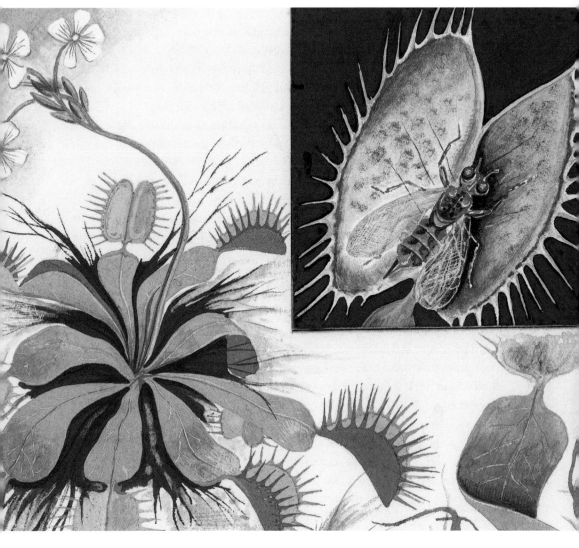

Killer plant: botanical drawing of Venus fly-trap.

But Darwin was fascinated by plants that disrupted the *scala naturae*. Plants that moved and plants that ate delighted him: in 1870, eleven years after the *Origin*, he published *Insectivorous Plants*, and inevitably the Venus fly-trap was one of the stars. He also discussed the sundews of the genus *Drosera*, which contains nearly 200 species. These are insectivorous plants that operate not with a snap-trap, like the Venus fly-trap, but with glue, on the flypaper principle. And he nailed it, as was his uncanny wont, noting that insectivorous plants tend to live in poor soil, with little nitrogen, where growth would be limited or impossible 'unless the plant has the power of obtaining this important element from captured insects'.

The evolutionary advantage of an insectivorous diet was made clear. Insect-eating has evolved at least six times among extant plants: a convergent evolution that is in itself an eloquent demonstration of natural section. Darwin then discussed the Venus fly-trap and suggested that the mechanism, which at first closes imperfectly, deliberately allowed small insects to escape, because the energy-expense of digestion is only worthwhile for more substantial prey.

Darwin's theory of evolution by natural selection is brilliantly argued in the *Origin*. In his later work he expands and adds further demonstrations of the way it operates: the way life on earth works. The Venus fly-trap is an attention-grabbing detail. The plant operates by a canny mechanism that requires the over-scrupulous writer to overload the text with inverted commas, explaining that the plant 'knows' and 'remembers'. The trap is triggered by contact with one of the three sensitive hairs – trichromes – that lie within each half of the plant's trap. But a single random event – a blown seed, a raindrop – will not trigger it. The trap will not operate unless another hair is triggered within the next 20 seconds, or the same hair is stimulated a second time. Then the trap closes.

The time this takes is variable, depending on light, humidity, size of prey and the conditions in which the plant is growing – but is around half a second. The teeth interlock, and if the prize is big enough to make the process worthwhile, the trap closes fully, seals itself and becomes a stomach, and the bright red digestive fluid gets to work. In short, the plant is aware that it has been contacted, aware of elapsed time, and can count the number of times it has been struck… and that allows it to move and to eat. The prey is digested in the course of the next week or so. The plant's prey is about one third ants and one third spiders; the rest is made up of beetles, grasshoppers and a few random others – including some small vertebrates like frogs. The closing mechanism is not fully understood; there are various theories, some of them to do with the rapid movement of calcium ions.

But why Venus? What has the goddess of love got to do with a carnivorous plant? The answer lies deep in the byways of Jungian archetypes and Freudian mythologies, which the reader is invited to deal with according to personal taste. Some have suggested that the name comes from the plant's pretty white flowers, as if the flowers were the most memorable thing about the plant. This is disingenuous: the fact is that the plant's traps remind people of female genitalia. The spikes surrounding the trap chime in with the ancient notion of the *vagina dentata*, the toothed vagina: an idea based on fear and distrust of female sexuality and female power. In researching this curious topic, I came upon a website that illustrated the *vagina dentata* with an unexpected bit of Photoshopping: a naked woman wearing not a fig leaf but, yes, a Venus fly-trap.

There is an American vernacular name for the plant, tipitiwitchet. This has been described as a Native American term, a claim that has been ridiculed elsewhere. A tippet is a fur collar; a twitch is a small noose used on the upper lips of horses as a calming device; the term is also used for traps for small mammals. Again, readers may make their own minds up; but it's worth noting in passing that all flowers are sexual organs. The traps from a Venus fly-trap are not flowers and they are not sexual in function or origin: they are modified leaves.

The fearsome plant is an idea of archetypal force. In 1874 a hoax about the man-eating tree of Madagascar was published in the magazine *New York World* and was widely believed. Edgar Rice Burroughs wrote about the Plant Men of Barsoom; in the Harry Potter stories, the young wizards are caught in a vicious plant called Devil's Snare. Such ideas can be cheerfully burlesqued: *Attack of the Killer Tomatoes* is a spoof horror film of 1978. The musical *Little Shop of Horrors* is based around a man-eating plant called Audrey II. In the television series *The Addams Family*, Morticia tends a lavish collection of house plants, and feeds one of them – Cleopatra – with large steaks.

But none of these can rival the triffid. In Wyndham's classic apocalyptic novel, triffids are semi-domesticated plants cultivated for the rich quality of their oil. They can move, and they carry deadly stings. The stings can be removed, but the oil is better when the plants are allowed to keep them. The novel begins with a meteor shower that leaves most of the world blind. In the chaos that follows, the triffids get loose and cause terror almost – almost? – as if they were acting with coordinated intelligence.

The Venus fly-trap grows to be around a foot tall (30cm) at best, with traps about an inch (2.5cm) wide. Not very threatening to human survival, then. But the threat they pose to human minds is immense. We like to be clear about things: plants are plants, animals are animals and humans are humans; back to the *scala naturae*. We like to think that we humans are a little like animals, but rather more like angels. A belief in God and angels is not, it seems, necessary to support such a view. Few people ever speak of such a concept out loud, but it's essential to the way we live and manage the world – and we don't like it to be disturbed. But at the same time, we find such disturbance spookily – and subversively – attractive. Venus fly-trap knows, Venus fly-trap remembers, Venus fly-trap eats. Venus fly-trap tells us that our unspoken but deeply felt ideas about the way the world should operate are frighteningly and thrillingly wrong.

TWENTY-EIGHT
YEW

My shroud of white, stuck all with yew,
O, prepare it!

William Shakespeare, *Twelfth Night*

For uncountable millennia the main thrust of human culture has been the search for an answer to the overwhelming question: what happens when we die? The question of what our ancestors believed about life has no straightforward answer: traditions die, no written records are left if they ever existed, and enigmatic objects and constructions can be interpreted in many different ways. Between cautious scholarship and wild speculation, we reach the conclusions our individual temperaments prefer.

But again and again it goes back to death and what happens next: rebirth, regeneration, the continuing nature of life. Our ancestors die but they survive in us. Every few months the plants that feed us die or stop producing food, but next year they come again. In many legends we can trace the still more ancient idea of the vegetation god: a god who represents the forces that govern the seasons: a god who, in dying, lives again. In the Arthurian legends a sick and impotent king rules over a dying land: but his kingdom and the king can be saved. All it takes is the Holy Grail and a knight to seek it. These mythologies were brought to the world's attention in a new form in 1922, when T. S. Eliot published his long poem *The Waste Land*, turning fertility myths upside down in an ecstasy of modernist despair:

April is the cruellest month, breeding
Lilacs out of the dead land, mixing
Memory and desire, stirring
Dull roots with spring rain.

And again and again our understanding of such ancient matters is tied up with the yew tree: a species planted in hundreds of churches across Britain and northern France, many of the trees older than the churches they stand beside, and all of them heavy with traditions older than Christianity. It is speculated that the yew trees mark places that were sacred to the local people long before Christianity was

Opposite Ancient worship: yew tree in an English churchyard, c.1890 (artist unknown).

established; that these places were taken over by Christianity without entirely displacing local customs and beliefs.

The yew is one of three conifer species widely accepted as native to Britain, along with Scots pine and juniper. It doesn't grow so very tall, about 33 feet (10 metres) is usual, with a maximum of about twice that. It is a conifer but, confusingly, its modified cones look like berries and are called arils. They are bright red, open at the top and taste sweet. Yews are famously long-lived, but an exact age is impossible to verify. That's partly because the wood that makes up the tree is unlikely to be as old as the tree itself: it operates a system of continuous replacement, and can be compared to the man who had his grandfather's axe – my father replaced the handle and I replaced the blade. The boughs become hollow so you can't count the annual growth rings. Those who seek an exact age are left with tradition, folklore and local patriotism: not the best route to objective truth. It's likely that most of the yews in country churchyards are around 500 years old, with some of them twice that. A few exceptional trees are even older.

With yews we come again on the tradition of naming individual trees: the most famous is the Fortingall Yew in Scotland, which is probably about 2,000 years old, perhaps the oldest tree in Europe. Local tradition gives a still greater age; up to 9,000 years has been claimed. In one story, Pontius Pilate was born beneath it.

Every part of a yew tree is poisonous apart from the berries, which get eaten by birds who disperse the seeds in their droppings. Pliny the Elder reported that 'the poison of the Arcadian variety has such an instantaneous effect that it is fatal to sleep under it or to eat one's food beneath it'. Death and life meet in these trees. It's been calculated that 2 ounces (50g) of needles are enough to kill an adult human: a pretty substantial helping. There is information about yews in Caesar's *Gallic Wars*; he wrote that Cativolcus, chief of the Eburones, a Gallic-Germanic tribe that lives in what is now the southern Netherlands, poisoned himself with yew rather than submit to Rome; perhaps the choice of suicide method was significant. The witches used yew as one of the many ingredients of their potion in *Macbeth*: 'Slips of yew, silvered in the moon's eclipse.'

There are vernacular traditions in every religion, varying from place to place, sanctioned by custom rather than liturgy. There was a tradition of placing a sprig of yew on the graves of the recent dead on All Saints Day, and of carrying yew sprigs in the annual procession of Palm Sunday; on Palm Sunday Christ rode into Jerusalem to be put to death, only to rise again. In the intensely seasonal world of the northern hemisphere above the tropics, the annual rhythm of the year brings death in the winter followed by rebirth in the spring. The trees that stay green stand for hope during the darkest months, as we have seen in Chapter 26 on Christmas trees. The yew, which not only stays green but apparently never dies, was a redoubled symbol of hope. Even if it does die, it has usually lowered some of its branches to the ground and these have taken root: a new tree that is still the old tree lives on: a dead thing reborn.

There is a more prosaic reason for the prevalence of yews in churchyards. The trees shed needles, twigs and branches in the natural course of things, and these are poisonous. It makes obvious sense, then, not to graze your livestock under its branches. The presence of a yew tree would stop any commoner from sneaking his livestock into a churchyard for a good graze.

The needles, branch tips and bark of yews have been used to treat an alarming variety of ailments, including diphtheria, tapeworm, epilepsy, rheumatism and infections of the urinary tract. More recently, yew species have played a significant part in the development of chemotherapy.

Yew wood is good to work with: the hardest of the softwoods. One of the world's oldest surviving wooden artefacts is the Clacton spear, made of yew and more than 400,000 years old. Yew wood has an extraordinary quality of elasticity: and that makes it the best wood for a longbow. A yew-wood bow is made with heartwood on the inside, for compression, and the sapwood outside, for tension. It takes a lot of tree to make one; the wood is often twisted and knotted and you need to discard a good deal.

The power of the yew-wood longbow played a crucial part in the Hundred Years War between England and France. The efficacy of the bow as a long-range weapon was shown dramatically in the 1944 film of Shakespeare's *Henry V*, directed by and starring Laurence Olivier; in its course the French troops are routed by the English archers at the Battle of Agincourt in 1415. The archers were still more effective at the Battle of Crécy in 1346, in which France lost eleven princes, 1,200 knights and 30,000 soldiers, while the English lost 100 men in total.

A yew bow is powerful thing. Under testing, an arrow has been driven 2 inches (5cm) into oak from 220 yards (200 metres). Pointed – as opposed to broad-headed – arrows could easily penetrate the plate armour worn by mounted soldiers. The problem with this super-weapon is that there is a finite supply of material to make it from, since yew trees require half a millennium to regenerate. This led to primordial conservation projects: Edward I ordered the planting of yews, but this is a long-term business. So yew wood and bows were imported; in 1472 it became law that every ship coming into an English port had to bring in four bow staves for every ton imported; later this was increased to ten. This import was also a primordial exercise in the export of conservation problems: a practice that the developed nations have enthusiastically followed ever since.

Thus the yew tree unites sacred and secular. The tree is associated with religion, both pagan and Christian, and with life, death and rebirth. It also hints at the superimposition of Christianity onto pagan customs and sites. But we also have the yew tree as a crucial item in the pursuit of worldly power. The ruler who could command the yew had the edge.

TWENTY-NINE
BARLEY

Corn rigs and barley rigs,
Corn rigs are bonnie!
I'll not forget that happy night
Among the rigs with Annie!

Robert Burns, 'The Rigs O' Barley'

They buried him, lay clods on his head, and when he stood up again they cut him off at the knee, rolled him and tied him, pricked him in the heart, split him skin to bone, ground him between two stones – and still they couldn't kill him, because next spring he would rise again. He's John Barleycorn:

There's beer in the barrel and brandy in the glass,
But little Sir John, with his nut brown bowl, proved the strongest man at last.

It's an old English folk song that comes in many versions, but at the heart of it all is the annual death and rebirth of the crop. Why barley, rather than any other crop? The first reason is that barley was one of the first grain plants to be cultivated; wild barley still grows all over the Fertile Crescent. It has been grown for at least 10,000 years; traces of barley have been found at pre-pottery sites; pottery was invented around 9,000 years ago. This is an ancient food.

'Barley is the oldest food,' wrote Pliny the Elder. It was thought to be strength-giving: gladiators ate a good deal of it and were, Pliny says, called 'barley men'. It could be eaten as flatbread and as porridge, filling and nutritious. But barley contains much less gluten than wheat, and therefore makes a poor leavened bread; the yeast needs the stickiness of the gluten if it is to rise. As a result wheat gradually took over as a staple in Europe.

Barley has many advantages as a crop, because it is remarkably tolerant. You can grow barley in sub-tropical and sub-Arctic climates; it can be grown to harvest in a mere ninety days. In medieval Europe barley was a peasant staple; wheat was for the wealthy. Barley was a vital European staple until potatoes (see Chapter 32) took over in the nineteenth century. Barley remains an important food in Arabic, Persian and Kurdish cuisines.

Opposite Drink more beer: advertisement by Marc-Auguste Bastard (1863-1926).

BIÈRES DE LA MEUSE

IMPRIMERIES LEMERCIER, 57, Rue de Seine, PARIS.

But barley is still grown in quantities. It has a long stiff bristle growing from the ear – John Barleycorn's beard, or, if you play Scrabble, the awn. Barley (in the United States, half the entire crop) is used as animal feed. Barley straw makes excellent bedding for livestock. But the main use of barley in the modern West is for drink. It's the base material for beer and whisky.

For this, the barley needs to be malted: that is to say, the seeds need to be separated from the plant and then germinated. They are steeped in water two or three times a day for two or three days, and that does the job. Germination changes the nature of the seeds, as you would expect, modifying starches into sugars and developing enzymes. This drastically improves the flavour and gives the yeast (see Chapter 11) more sugar to work on when it comes to fermentation.

Once the seeds have germinated they must be dried. Traditionally they were dried in the air for five days in specialised buildings known, obviously enough, as maltings. After that the fermentation can begin... though there are non-alcoholic products made from malted barley, including malt vinegar, a harsh-tasting vinegar traditionally used on fish and chips in Britain, and night-time drinks like Horlicks, Ovaltine and Milo. Malt extract was used as a dietary supplement in the first half of the twentieth century, to make up for the nutritional poverty of the urban working-class diet. In *The House at Pooh Corner*, the second Winne-the-Pooh book, Tigger moves into Kanga's house 'and had Extract of Malt for breakfast, dinner and tea'.

But most barley grown for humans is used to make alcoholic drinks. Under the famous German purity laws that date back to 1516, the only permitted ingredients in beer are water, barley and hops; the yeast being understood. For Europeans, barley is the stuff you make beer from. In Scotland and Ireland, it's also the stuff you make whisky from.

You can buy whisky in two forms: blended and single malt. A blended whisky usually has a base of neutral grain spirit: that is to say, a very pure form of hooch, stuff with no real taste but a powerful kick. This is mixed, in varying proportions, with malt whisky, which is distilled from malted barley and then stored in wooden casks. The casks give flavour and are open to the air, allowing the spirit to mature and become pleasant to drink. (You can keep whisky 100 years in a sealed glass container and no alteration will take place.) Blends differ in their degree of sophistication: the more malt whisky they contain, the more palatable the drink. Some very fine blends, with little or no neutral grain spirit, are more expensive (and better) than run-of-the-mill single malts.

A single malt is the product of a single distillery, and contains no neutral grain spirit. To acquire this status, the whisky must be pot-distilled; that is to say, with a device that requires the spirit to be distilled in batches, rather than as a continuous process. The local water is a crucial ingredient, sometimes for its purity, sometimes for the additional flavours it brings.

The name whisky is from the Gaelic: *uisce beatha* is the 'water of life'. (Scotch whisky is spelled without an 'e', Irish and American usage prefers whiskey. Japan, which produces increasingly excellent whisky, prefers the Scottish spelling.) And that brings us back to John Barleycorn: for the song, though connected to the cereal crop that forms a staple diet, also references drink.

There's beer in the barrel, and brandy in the glass...

Though surely that would sing better as whisky. George Bernard Shaw was intrigued by the song and mentioned it in his preface to *Androcles and the Lion* (which is longer than the play). 'And from the song John Barleycorn you may learn how the miracle of the seeds, the growth, and the harvest, still the most wonderful of all miracles and as inexplicable as ever, taught the primitive husbandmen... that God is the seed and that God is immortal.'

The point was taken up by C. S. Lewis, who suggested that the fertility cults of death and rebirth were not forerunners of Christianity, but its anticipation: 'I think the thrill of the pagan stories and of romance may be due to the fact that they are the beginnings – the first faint whisper of the wind from beyond the world – while Christianity is the thing itself.'

Again, readers may make up their own minds according to belief and temperament. But let us leave this chapter with one last song about death, resurrection and the products of barley. It concerns Tim Finnegan, an Irish hod-carrier who, unstable after the previous night's drinking, falls off his ladder and breaks his skull. So they hold a wake for him, a fight breaks out, whiskey is spilled over the corpse – which instantly revives:

Whirl your whiskey round like blazes
Thundering Jesus, do you think I'm dead?

James Joyce loved the song, and put its themes of life, death and renewal into his last mad masterpiece, *Finnegans Wake*. He dropped the apostrophe for his love of ambiguity: in the course of the book the Finnegans do indeed wake up. You can regard this as the greatest literary masterpiece ever inspired by a cereal crop – though you might prefer a haiku of Bashō, the eighteenth-century Japanese poet.

Girl cat, so
thin on love
and barley.

THIRTY
BODHI TREE

It is better to conquer yourself than to win 1,000 battles.

Gautama Buddha

Death is not the end. Every plant in this collection tells you that without ambiguity: for each one that dies will, if all has gone well, have left behind descendants. The individual plant is no more, but its immortal genes live on. You can argue that all plants – not just barley and yew – teach us the meaning of life. But it has been the pattern throughout history and across cultures to revere some plants above others: for their beauty and their meaning.

The Bodhi tree is central to Buddhism. It was beneath the Bodhi tree in Bodh Gaya, Bihar, in India, that Siddhartha Gautama sat and meditated for seven weeks, found his own understanding of life's meaning and arose as the Buddha. The word *Bodhi* means enlightenment. Here, then, we have another tree of wisdom, another tree of knowledge: the difference here is that the wisdom was attained by the person's own efforts rather than any magical quality of the tree. The Bodhi tree is not an embodiment of mystical power but, if you like, an emblem of hard-won common sense: forget the romantic Western idea of enlightenment as a form of magic from the mysterious East.

The term Bodhi tree can be used of that individual tree in Bihar, of several other holy trees across Asia and the world, or of the species *Ficus religiosa*. It is also called the bo tree, the pippala, the ashwattha and the peepul. It belongs to the same genus as the strangler figs of Chapter 1 and throws the same inviting shade. Even in warm countries we don't appreciate shade trees as much as we used to, now we have buildings and air-conditioning. But in Asia in the fifth century BC, only a very rich and powerful person could own a shade as good as that of a fig tree. Anyone planning a long stay in a single spot would choose just such a tree.

Ficus religiosa is not a strangler; it works more subtly. A successful seed germinates on another tree, but instead of strangling from the outside, its roots penetrate the stem of the support plant and, working from the inside, eventually split it from within. At this point the tree becomes self-supporting and throws its branches out wide. It can reach around 100 feet (30 metres) in height with a trunk diameter of 10 feet (3 metres). They grow best on alluvial sandy soil with good drainage, and thrive in warm, moist climates. They are found all over the Indian subcontinent

Holy place: women at worship beneath the Bodhi tree; Indian miniature, eighteenth century.

and Indo-China, and have been introduced in many places across the world, sometimes for their religious power and sometimes for their decorative value.

The Bodhi tree of Bihar is important in Buddhism because it is the completion of the story. The prince Gautama Siddhartha lived in great luxury, but when he went outside the royal compound he discovered illness and death. When he understood the inevitability of suffering, he took up a life of extreme asceticism, but eventually decided that this was not the answer. He accepted the gift of rice pudding (see Chapter 17), realising that the best path was the Middle Way, avoiding extremes. Then he sat beneath the Bodhi tree and meditated: on the necessity to be freed from greed, hate and delusion, and on the importance of the Four Noble Truths: of suffering, of the cause of suffering, of the end of suffering and of the path that leads to the end of suffering.

Buddhism was established on the basis of the Buddha's enlightenment – Gautama's breakthrough meant that he became a Buddha, traditionally depicted in the position of meditation. The place of his enlightenment became a pilgrimage site, but the tree itself has had a troubled history. The Emperor Ashoka ruled most of the Indian subcontinent in the third century BC, converted to Buddhism and paid homage to the tree. His second wife Tishyaraksita didn't care for this, and had the tree poisoned with mandu thorns. But the tree survived, and Ashoka caused a 10-foot (3-metre) wall to be built around it to keep it safe. But fifty years later, King Pushyamitra Shunga destroyed the tree as part of his campaign against Buddhism. The tree was replaced; in the seventh century AD it was destroyed by King Shashanka. It is a centuries-long war against symbols, revealing fierce intolerance of a religion that preaches tolerance. In 1876 the tree that had been planted in the place of all those fallen ones was destroyed in a storm; the present tree was planted by a British archaeologist, Alexander Cunningham, in 1881.

If you are a Buddhist looking for a tree with an ancient history of devotion, go to Sri Lanka, where the Jaya Sri Maha Bodhi has been growing since the third century BC. It is reckoned to be the world's oldest historical tree of religious significance. The original sapling was brought there by the nun Sanghamitta under the patronage of Ashoka; she is sometimes said to be his daughter. The tree stands in the island's ancient capital Anuradhapura. Cuttings from the tree have been cultivated at Buddhist sites around the world.

The day of the Buddha's enlightenment is celebrated every year on 8 December. It is often marked with the giving and eating of heart-shaped biscuits: biscuits in the shape of a leaf of the Bodhi tree (cordate to a botanist). It is also appropriate to feast on rice pudding, a luxury for poor people, and to greet people with the words *budu sarani*: may the peace of Buddha be with you.

The imposing and inviting fig tree is also important to Hindus. In the *Bhagavad Gita*, Lord Krishna declares: 'Of all the trees I am the peepul.' A peepul tree is a natural place for sadhus – Hindu ascetics – to meditate. Such trees are also suitable for a *pradakshina*, a sort of circumambulatory prayer or meditation: you walk around the tree chanting *vriksha rajaya namah*, or salutation to the king of trees.

Buddhism teaches the desirability of moving beyond life. In the Fire Sermon, the Buddha preached: 'He becomes divested of passion, and by the absence of passion he becomes free; and he knows that rebirth is exhausted, that he has lived the holy life, that he has done what it behoved him to do, and that he is no more for this world.' Across the world, the fig trees bloom and fruit and seed, inviting travellers to sit for a while in the coolness of their shade.

THIRTY-ONE
MAGIC MUSHROOMS

Turn off your mind, relax and float downstream

John Lennon and Paul McCartney

Several chapters in this book are about plants and fungi that alter the consciousness of humans who consume them. The human need to be something other is answered by a number of obliging species which make this possible. A good few species of fungi can give the consumer a violent, intense experience quite unlike everyday living.

At least 200 species of fungi contain the hallucinogenic substances psilocybin and psilocin, and many are not closely related. We will also consider the fungus ergot, from which lysergic acid diethylamide – LSD – was first synthesised in 1938. A plant or fungus that develops a dangerous substance usually does so for self-protection: in order not to get eaten. It has been speculated that the dangerous chemicals in hallucinogenic fungi suppress the appetites of predatory insects. In other words, we are invited to accept the notion that a fungus is manipulating the minds – the consciousness – of insects to its own advantage. That's certainly a trippy idea.

But the deliberate consumption of hallucinogens by humans in search of a heightened experience predates the hippies by at least half a dozen millennia. There are claims that mushroom use is shown by rock art in Europe and North Africa up to 9,000 years old. The use of mushrooms is much more explicit in the later art of pre-Columbian Central and South America. A small sculpture that clearly depicts the mushroom *Psilontainxicana* was found in West Mexico and dated to AD 200.

It is generally assumed that hallucinogenic mushrooms were used for communion, divination and healing. The idea in all such experiences is to get closer to God: to escape the normal drudgeries and cares of existence while seeking something beyond. Hallucinogenic mushrooms were part of the feast that commemorated the coronation of the Aztec King Montezuma II in 1502; mushrooms were important on many ritual occasions. The sixteenth-century Spanish priest Bernardino de Sahagún reported a banquet: 'The first thing to be eaten at the feast were small black mushrooms that... bring on drunkenness, hallucinations and even lechery; they ate these before the dawn... with honey, and when they began to feel the effects, they began to dance, some sang and others wept… when the drunkenness of the mushrooms had passed by, they spoke with one another about the visions they had seen.'

The eagerness with which hallucinogens have been sought has been matched throughout history by the eagerness of others to suppress them. Catholic missionaries in South America considered the indigenous religion idolatry and the use of mushrooms a way of communicating with devils; consequently their use was forbidden. All the same, it's hard to outlaw things that grow wild. In 2005 the British government made it illegal to be in possession of any part of the liberty cap mushroom *Psilocybe semilanceata*, which is all very well, but who is responsible for a colony growing on the local cricket pitch?

It has been speculated that European hallucinogenic fungi were used in witchcraft, allowing the users to experience flight, to travel in the bodies of non-human animals, to become shapeshifters and so forth. The history of witchcraft is shadowy and elusive, and much pursued by people with an agenda of one kind or another. What remains fact is that where there is knowledge of hallucinatory substances, there will be people eager to use them.

There will also be people who ingest them by accident. In 1799 Dr Everard Brande submitted a paper to the *London Medical and Physical Journal* describing a case in which a man collected mushrooms in Green Park, London and served them up to his family. Alas, they were liberty caps. He reported that 'the youngest child was attacked with fits of immoderate laughter, nor could the threats of his mother and father refrain him'.

Ergot is another hallucinogenic fungus, unrelated to any of the psilocybin mushrooms, and occurring as mould on some species of grass, notably rye. This can be dangerous, even deadly when baked in bread. Gangrenous and convulsive forms of ergotism exist. An eighteenth-century outbreak in France caused 8,000 deaths. The substance also causes extreme hallucinatory experiences. It has been suggested, fairly convincingly, that ergot poisoning was behind the outbreak of mass hysteria known as the Salem Witchcraft Trials, which took place in 1692-93, in which more than 200 people were accused of witchcraft, thirty were found guilty, nineteen were executed and at least another five died in jail. People on both sides of the argument were afflicted; research showed that the weather that year was damp for sustained periods, ideal conditions for the mould to flourish on stored grain.

The fascination with hallucinogenic mushrooms dates to the middle of the last century. In 1955 Gordon Wasson and his wife Valentina Pavlovna took part in a mushroom ceremony in Mexico. They wrote this up for *Life* magazine in 1957 and caused quite a stir; a year later Albert Hoffmann, a Swiss chemist, identified psilocybin and psilocin as the active ingredients of the mushrooms. Timothy Leary, a Harvard psychologist, was inspired by this, and went to Mexico in search of

An Aztec feast: from the sixteenth century Florentine Codex *by Bernardino de Sahagún (c.1499-1590), depicting the smiling faces of sacrificial victims consuming mushrooms before decapitation.*

a similar experience. In 1960 he and Richard Alpert established the Harvard Psilocybin Project.

Hoffmann had investigated the properties of ergot in 1938 while seeking a cure for migraine, and he isolated lysergic acid diethylamide, or LSD. He discovered its hallucinogenic properties in 1943, when he accidentally ingested some. The substance was then tested by the CIA, on the grounds that it might be useful for mind control. People were tested for their response to the drug, some taking it without their knowledge. The substance got out, and its extraordinary properties defined a generation. LSD was manufactured and used informally by its consumers for spiritual and psychological development. It was a new gateway to righteousness.

Leary's project included the Concord Prison Experiment, in which he claimed that prisoners vowed to reform their lives after taking guided trips with LSD. He also said that LSD 'cured' homosexuality. He took the drug himself, and encouraged (or pressured) his students to do the same. The project was closed down and he was fired in 1963. He then became a proselytiser for LSD, coining the phrase 'turn on, tune in, drop out'. President Richard Nixon described him as 'the most dangerous man in America'. Besides preaching love and peace, he was also a major attention-seeker. His books, notably *The Politics of Ecstasy*, were as important on the bookshelves of educated hippies as the *Bhagavad Gita*, and as little read.

Ian MacDonald, in his excellent book about the songs of the Beatles' *Revolution in the Head*, wrote of beliefs current in the late '60s: 'With LSD, humanity could transcend its "primitive state of neurotic irresponsibility" and, realising the oneness of all creation, proceed directly to utopia'. The understandings gained by its use were to make better individuals, who would create a new society by way of a new perception of reality. They were glorious times that produced masterpieces in many different genres. There were also many casualties, people who lost their way under the relentless assault of the drug: 'Russian roulette played with one's mind', as MacDonald wrote. A great friend of mine announced one evening that he was a genius. 'Do you have an infinite capacity for taking pains then?' he was asked. 'No – but I've got an infinite capacity for taking drugs.' How we laughed.

Meanwhile, the fungi still grow. The dangers of contracting ergotism were lessened as early as the nineteenth century by better draining of fields and better cleaning of the grain, and also by a shift to potatoes as a staple. Psilocybin mushrooms grow as they have always done, and those with a taste for such things seek them out across the world. What has changed is the widespread perception that such experiences have something magical about them, that they are a direct way of encountering God. Few people suggest that LSD is a sacrament these days. As John Lennon (nominally in partnership with Paul McCartney) wrote in his song 'Tomorrow Never Knows':

Lay down all thoughts, surrender to the void
It is shining
It is shining

THIRTY-TWO
POTATO

You like potato and I like pot-ah-to.

Ira Gershwin

A tuber is a thick stem, modified for the underground storage of nutrients; it can also form buds that allow the plant to reproduce asexually. A plant that develops tubers is investing for the future: it's aiming to survive the winter and use the stored energy to grow again, a process that repeats itself year after year. It follows that a tuber is a good food source for organisms other than the plant that went to the trouble of growing it. These tubers stay in the soil for month after month, available even when the growing season is over. If you can access palatable tubers, you will survive beyond the next day. That is exactly what a potato is: an accessible and palatable tuber, packed with energy and relatively easy – and therefore cheap – to produce. It was the potato that made mass nutrition possible: and that changed the world.

It's an unexpected achievement for a plant that belongs in the same family as deadly nightshade and is reliably poisonous in its shoots, flowers, fruits and sometimes in the skins of the tubers themselves. The fruit of the potato plant are like cherry tomatoes – potatoes and tomatoes are related – and contain around 300 seeds, which are also toxic. But when they're not exposed to light and turning green, the tubers are supremely palatable. There are around 150 wild species of potatoes, growing in South and Central America. They were probably domesticated several times over. They were the main energy source for the Inca Empire, and were a pre-Columbian staple in many places.

Potatoes, in the form of the species *Solanum tuberosum*, were brought to Europe by explorers towards the end of the sixteenth century. They reached Spain around 1570 and Britain a couple of decades later; the credit is sometimes given to an employee of Sir Walter Raleigh, the polymathic Thomas Harriot, and sometimes to Sir Francis Drake. The potato is not closely related to the sweet potato, and in early records is differentiated by name, and called the bastard, Virginia and white potato.

It was not an instant change of staple: humans are conservative about what they eat. Potatoes were cultivated, sometimes as an exoticism, sometimes even for the beauty of their flowers, which can be pink, white, red, blue and purple. They grow well in Northern Europe, where it is moist enough for the tubers to thrive.

But there's an obvious disadvantage: they don't store well, nothing like as well as grain, which can last several years if kept appropriately clean and dry. Even carefully stored potatoes can rot.

It was also impossible in many places to grow potatoes in open fields: long-standing systems of agriculture involved integrated practices like the grazing of stubble (what is left on the field after wheat, barley and rye have been harvested). The mass growing of potatoes would disrupt such systems, not least because their foliage is toxic. Potatoes were somewhat mistrusted, called the devil's apple, because they grow underground and must be stored in the dark if they are not to sprout. But changing circumstances made the potatoes more desirable.

Potatoes have certain advantages over grain, though. One is that they are less easy to pillage. That makes them a safer staple in troubled times; potatoes are much more bulky than grain or flour, and therefore much harder to steal in quantity. Potatoes already have their own water inside them, but water must be added to grain to make bread or porridge. The Little Ice Age, the fluctuation in climate (not a true ice age at all) that brought lower temperatures, conventionally dated from the mid-seventeenth century to the mid-nineteenth, meant that potatoes were easier to grow than grain. They are also more economical of space: you can get between two and four times the amount of energy per acre from potatoes than from grain.

Frederick the Great of Prussia was a great champion of the potato, and in 1756 passed the *Kartoffelbefehl*, or potato order, to encourage their growth. The idea was to save his people from famine, but there was inevitable resistance to this: no one wanted to eat the damn things. The story is that Frederick grew potatoes and guarded them heavily because they were now *the* royal vegetable. But the guards were phonies. They had been ordered to tolerate any thefts that might take place. So potatoes, stolen rather than given or enforced, were grown in secret and became, in every sense, an underground success. The potato found another champion in Louis XVI of France; his wife Marie-Antoinette attended a ball in a headdress of potato flowers. In 1838-39 Russia suffered grain failure and, with it, the beginnings of a conversion to potatoes. By the nineteenth century potatoes were an important staple in much of Europe.

Potatoes went to North America with the colonists; the species made a double Atlantic crossing. They were becoming increasingly important to human nutrition: easy to grow, immensely filling and, above all, cheap. They need little labour, and you can eat them straight out of the ground, rather than sending them to the mill and then using the resulting flour to make bread. As a result, more people could stay alive: what ecologists call the carrying capacity of the land (or of the society) increased with the spread of the potatoes. A rising population made it possible for the process of industrialisation to take place: generations of potato-fed miners

Keeping body and soul together: The Potato Eaters, 1885, by Vincent van Gogh (1853-90).

took coal (see Chapter 16) from the earth to supply the energy needed by the industrialising society. This must not be confused with a recipe for instant happiness. Vincent van Gogh (see Chapter 8 on sunflowers) painted his harrowing picture of *The Potato Eaters* in 1885: a for-all-time image of the dirt poor, in which desperation is mingled with a heroic willingness to endure.

The adoption of the potato allowed the population of Ireland to expand, though still in desperate poverty. An acre could provide enough potatoes to last a family for a year, along with the milk from a single cow and the meat from a pig that could be fed on scraps. The bulk of the population depended on a single variety of potato, the Irish Lumper. Monocultures are not by nature resilient: and so when potato blight struck Ireland in 1845 it was a disaster. Famine lasted four years;

in that time 1 million people died and 1 million more emigrated, some to England, many others to North America.

Potatoes have become a major food source all over the world: the fourth most widely grown crop after rice, wheat and maize. In theory they provide excellent nutrition and bulk with comparatively low calories. The problem is the potato's extraordinary affinity for fat. Every way of making potatoes more delicious involves fat: oil or butter for frying, butter, cream and cheese for mixing. Reckless use of potatoes cooked in this way increases the risk of obesity, diabetes and heart disease. The deep-frying of potatoes at home is potentially dangerous and requires investment in oil or fat, and so it became a commercial operation in poor districts of the UK.

After the middle of the nineteenth century it was possible to buy fish and chips in many working-class districts: potatoes cut into lengths, deep-fried and served with fish fried in thick batter: a cheap, tasty and filling treat. You could take it away, wrapped in newspaper, and eat it at home or in the street: take-away food was invented. George Orwell, in his survey *The Road to Wigan Pier*, suggested that the existence of fish-and-chip shops helped to avert revolution: by the 1930s Britain had 35,000 fish-and-chip shops. It has been speculated that the chips were at first used not as an accompaniment but as an alternative to fish: when the catch failed, people fried potatoes cut into fish shapes.

Potatoes are also used to make alcoholic drinks: vodka in Russia, akvavit in Scandinavia, and poitín in Ireland. The illicitly distilled poitín of Ireland is sometimes deadly, but when well prepared, it is delicious enough, with a pleasant echo of baked potatoes in the taste. I once savoured the contents of a very acceptable bottle that had been smuggled from Ireland to England by a nun as a gift to a priest.

We have also found a new way of consuming potatoes in what the British called crisps and the Americans chips. There are several claims for the origin of these. Root vegetables, sliced thin and fried, are traditionally served with pheasant and venison in Britain, and called game chips. There is the legend of Saratoga Springs, in which a customer was tormenting the cook George Crum by constantly sending back the potatoes he was served: too thick, too soggy, insufficiently salty. Crum, an African American, grew exasperated, sliced the potatoes thin to the point of transparency, fried them rigid and then dowsed them in salt. That would show him – but the customer thought they were delicious. In 1910 Mikesell's Potato Chips Company was operating in the United States. I can remember eating Smith's Crisps in England in the 1950s: the salt was contained in a screw of blue greaseproof paper.

Potatoes are associated with the fast food or junk food industry: served with fried chicken or burgers. There is great stress on the uniformity of such products.

McDonald's fries are made mostly from a single variety of potato, the Russet Burbank. Their demands bring us back to the problems of monoculture: if you want to stop the potatoes rotting underground, you must make heavy use of fungicide. This also affects the land all around, with the phenomenon called pesticide drift.

Potatoes can be cooked to please the wealthy, with truffle oil, or as fondant potatoes, cooked slowly in stock and a great deal of butter. But potatoes have always fed the poor. In the nineteenth century they kept people alive, for often there was quite literally nothing else. Now in much of the developed world potatoes are associated with the problem of obesity among the poor. Once only rich people could afford to be fat; now, it seems, only rich people can afford to be thin. Either way, potatoes are the plants of poverty.

THIRTY-THREE
TOADFLAX

Anyone not specially interested in the subject need not attempt to
read all the details.

Charles Darwin

Common toadflax is an avid coloniser of broken ground, establishing itself in patches and creating a floral display that looks like a cultivated garden. That's because the blooms are relatively extravagant for a wild flower, and also because they greatly resemble the related and much cultivated garden plant called snapdragon or antirrhinum. These flowers are or were popular with children because they open when squeezed, so you can make the flowers talk. Toadflax can be found on open ground across Europe and Central Asia into Siberia. They are attractive to many insect species, and that makes them important for roadside verges and field margins, linking up wild areas so that such places are not islands with inbred and therefore vulnerable populations. Toadflax has also played a significant role in our understanding of the entire concept of inbreeding.

The plant was important to both Linnaeus and Charles Darwin. Linnaeus called it a monster; for Darwin it was a classic example of the good luck that can come to a scientist who has the wit to notice it. Toadflax links these two scientists across a century and more of changing understandings of life.

Linnaeus, an outstanding observer of detail, knew all about toadflax. In 1742, a neighbouring botanist from Uppsala in Sweden discovered a colony of toadflax that weren't quite right; they had five spurs instead of one. This was, to a thinker with a good eye for a plant and a clear mind, the most shattering discovery. Most people would walk past the plants day after day without noticing anything unusual, but for a specialist it was startling. The shock of the discovery can be read in Linnaeus's response. Though he was normally cautious in what he set down on paper, he wrote: 'This is certainly no less remarkable than if a cow were to give birth to a calf with a wolf's head.' Linnaeus then went still further and called the aberrant plants peloria, which, from the Greek, means monster.

It was a disturbing discovery because the plants were asking the question that dare not speak its name. It was clearly a toadflax; equally clearly, it was not a toadflax

Opposite Eureka plant: yellow toadflax illustration by Margaret W. Tarrant (1888-1959), c.1930s.

Margaret W. Tarrant

as the world knew it. What did it mean? Linnaeus didn't know, but he certainly knew what it suggested. The implication of the rogue toadflax was that a species could change, that a species could become different, that a species could become something else – so much so that it might become another species altogether. The orthodox eighteenth-century view, maintained by both religion and science, was that a species was a species and would remain so, unchanged and unchanging, until the end of time. This was the period of the Enlightenment, the time of the protracted and in many ways still uncompleted divorce of science and religion. The toadflax, as humble a plant as ever brushed a walker's ankles or welcomed a bee, was asking some pretty enormous questions.

Linnaeus's view was that it must be some weird hybrid, the result of a freak breeding with an unknown other. This was wrong: but there is nothing wrong in being wrong: only in falsifying your data. Linnaeus grew toadflax in his summer residence at Hammarby and the plant continued to haunt his imagination. It remains as a sort of floral question-mark floating over his entire and magnificent oeuvre, the *Systema Naturae*, which went through twelve editions under his care and listed around 13,000 species in total. All the species that lived were there within its pages – or as near as he could get it, for his aim, naturally doomed to failure, was completion. But what if all these species were not, after all, immutable? What if life was not static at all, but dynamic: ever-changing: forever renewing?

Darwin, as we have already seen, provided the answer to that, not by inventing the concept of the mutability of species, but by showing us how it worked. His notion of natural selection is simple, almost blindingly obvious, once explained. As his great champion Thomas Huxley said after reading *On the Origin of Species*: 'How extremely stupid not to have thought of that.'

All of Darwin's considerable work after the publication of the *Origin* in 1859 reinforced the potency – the inevitability – of his original idea. So it follows that his 1876 book was another powerful statement – though it was, of course, delivered in his usual calm, almost apologetic tone – about the way that natural selection governs our lives and all lives. This book was *The Effects of Cross and Self Fertilisation on the Vegetable Kingdom*, a book so filled with detail that Darwin felt compelled to apologise for it. You might say it was a book from an original idea by *Linaria vulgaris*: the common toadflax, sometimes called butter and eggs.

Darwin is celebrated as a great thinker, and so he was. But he was also a great doer. He did stuff all the time: tried ideas out, set up experiments, never mind how long they were going to take. He sought the right results, not quick results. He did things and thought about them, and then tried more things, to see if they confirmed or if, on the other hand, they contradicted. Darwin loved to speculate: but he loved above all else to speculate on the data he had collected. He built his thinking on his data, not the other way round.

Which brings us to toadflax. He wrote: 'For the sake of determining certain points with respect to inheritance, and without any thought of the effects of close interbreeding, I raised close together two large beds of self-fertilised and crossed seedlings from the same plant of *Linaria vulgaris*. To my surprise, the crossed plants when fully grown were plainly taller and more vigorous that the self-fertilised ones.'

Here was a chance discovery: and as Alexander Fleming showed with his discovery of penicillin (see Chapter 21), chance can be the portal to discovery. Darwin realised that if plants try so very hard to avoid fertilising themselves, there must be a good reason for it: nothing evolves for no reason. He set up a series of experiments lasting eleven years, examining the differences between self- and cross-pollinated plants. He wrote: 'Cross fertilisation is sometimes ensured by the sexes being separated, and in a large number of cases, by the pollen [male] and stigma [female] of the same flower being matured at different times.'

He discovered that self-fertilisation isn't just an enhanced problem that comes with a weak parent propagating itself. Self-fertilisation is a bad thing even for a healthy plant: it is a bad thing *in itself*. Darwin was moving towards the discovery of the recessive gene. In a large population, recessive genes tend to remain undetected because they very rarely end up in the same genome: and it's only when that happens that problems can arise. But when you are breeding from closely related individuals – or still more drastically, from the same individual, acting the part of both parents – then the activation of recessive genes becomes almost inevitable. The importance of cross-fertilisation was an important discovery for everyone involved in the cultivation of plants: whether farmers or gardeners.

Darwin worked on many species of plants in the course of his experiments, including peas. But it was left to Gregor Mendel – as mentioned in Chapter 4 on peas – to take all this further and to come to the first clear understanding of genetics. And that, for good or ill, is the branch of science that most consistently looks to a radically different future for all of life on earth.

THIRTY-FOUR
CINNAMOMUM

*Make haste, my beloved, and be thou like to a roe or to a young
hart upon the mountain of spices.*

Song of Solomon 8:14

The genus *Cinnamomum* comprises 250 or so species of evergreen trees from
the family of *Lauraceae*, or laurels; they are mostly found in Asia. A couple of
these changed the world. *Cinnamomum verum* and *Cinnamomum cassia* are the
species from which cinnamon is mostly made: the sweet aromatic spice familiar to
most of us. You get it from the bark. The trees are cultivated by repeated coppicing;
you cut the tree down to a stump, from which a number of shoots emerge, grow
and are harvested. To make the spice you scrape off the outer bark, beat what's left
with a hammer and then remove the inner bark, which is now very thin. You roll
this up: and after a few hours in a dry, well-ventilated place you have cinnamon
quills. They can be used as they are or ground into a powder.

So far so humdrum. Spices are familiar and accessible in most cultures. It's hard
to believe that cinnamon and the other spices sold on the supermarket shelves
were once the most desirable commodities in creation and that the world was
turned upside down in pursuit of them.

Why? I was taught that it was because spices preserved food: put spices into
food and it keeps longer. Spices were therefore essential to survival in the days
before refrigeration. But that's not actually true. You can marinate a piece of meat
with all the cinnamon in your cupboard, but it will still deteriorate at the same
speed as the piece you didn't spice.

There's a second suggestion: spices disguise the taste of dodgy food, especially meat,
and so make it palatable. That's another doubtful proposition. You can cook rotten
food with all the spices you have left on the shelf after your cinnamon marinade and
it will still taste rotten. Spices don't hide flavour; they add flavour. Besides, if spices
did disguise the taste of gone-off food, it would be disastrous: bad food makes us ill,
sometimes terminally. We find decayed food disgusting because our bodies are warning
us not to ingest it. We need to taste the rottenness of bad food so that we can reject
it: it's an important survival mechanism and, fortunately, no spice can override it.

Opposite *Cinnamon plant: early nineteenth-century Chinese watercolour.*

flower.

I saw it at Luqua's in 1756. The plant
is common at Canton, the flower is ve.
sweet.

What spices do is make dull food interesting. Before refrigeration, fresh food was not something that could be guaranteed, even by the elite. Different kinds of food can be preserved by salting, drying, smoking and pickling, but most of such foods are better for being livened up. In winter, not much grows; in summer food goes off quickly. But the monotony of diets before refrigeration and imported delicacies could be relieved by spices. Spices traditionally make eating a pleasure as well as a necessity. That's true today: even the most conservative Western diets include tomato ketchup, which usually contains cumin, allspice, cinnamon and mustard, while Kentucky Fried Chicken has paprika, black pepper and mustard among its famously secret ingredients.

Spices are good because they increase our pleasure in eating. But spices were not always cheap and easy to come by. Eating and offering spiced food was a pleasure: but, crucially, it was an expensive pleasure. It was only available to the rich and powerful. It follows that spice proved your status: spice showed the world that you were better than the rest. Until recently spices were a luxury item in Europe, and the demand for this status-conferring treat was so intense that it drove the development of Europe and the world.

The desired spices were mostly grown in southern and eastern Asia, and they reached Europe through the Arab traders who mostly conveyed the stuff overland. Part of the trade was conducted along the most famous land-trading route of them all, which was developed for another commodity that conferred both status and luxury: the Silk Road (see Chapter 64 on mulberry).

Spices were important to Ancient Egypt at least 4,000 years ago. Often they were used for sacred purposes: burned in temples, used for embalming corpses. Sweet smells were important to a world without modern sanitation: the rich could separate themselves from the stinking commoners by breathing sweeter air. Spices were used in Ancient Greece and then Rome. Sappho wrote about the deliciousness of cinnamon in her poems of the seventh century BC; Nero burned a year's supply of cinnamon at the funeral of his wife Poppaea Sabina in AD 65.

The prices of spices were kept high by small quantities and extravagant tales about the fabulous difficulty of obtaining them. One such story concerned the cinnamologus, a ferocious bird that makes a nest of cinnamon sticks. They could be tempted away from these treasure-houses with lumps of meat: they devoured the meat, but when they returned heavy-laden, their nests collapsed, giving intrepid spice-gatherers the opportunity to collect the fallen sticks. Pliny the Elder reports the story and adds: 'There is also a story of cassia growing round marshes and protected by a terrible species of bats that guard it with their claws and by winged serpents. The locals inflate the price of the spices by concocting these tall stories.'

Spices were intensely desirable in Europe and there was no source but the trade in the eastern Mediterranean. Those that controlled the trade controlled the economy of Europe: and that's precisely what the merchants of Venice did from around AD 1000. They established a monopoly on the spice trade in Europe: the sole entrepot between east and west. Venice became a place of legendary power, dedicated primarily to economic rather than military might. It was a new kind of empire. By the late thirteenth century Venice had a fleet of 3,300 ships employing 36,000 sailors.

'Now, what news on the Rialto?' asks a character in *The Merchant of Venice*: he is referring to a bridge that crosses the Grand Canal, and does so in the way one might ask today about the news from Wall Street. The decline of Venice began with the growing power and ambition of the Ottoman Empire, and in the ambition of Europeans to the west of Venice. And still spice was at the heart of it.

Spices come in many forms: bark for cinnamon, the pit of a fruit for nutmeg, scrapings from that pit for mace, rhizomes for ginger and turmeric, flower buds for cloves, seeds for coriander, cumin, cardamom – and especially, black pepper. Black peppercorns were stuffed into the nostrils of the pharaoh Rameses II as part of the process of mummification.

With the spice trade under the control of the Ottoman Empire, Western Europe grew restless: the expansionist ambitions of the Ottomans were clear for all to see, and the people were not even Christian. There was unease at dealing with them: and besides, with advances in shipbuilding and navigation, the trade could be taken out of their hands by those with stout hearts and sound ships.

Thus the age of European exploration began: and its prime motivation was not the betterment of humankind or the quest for pure knowledge: it was spice. A new route to the lands of spice would bring riches beyond easy computation. Portugal, under King Henry the Navigator, was the first European power to succeed in this venture. In 1488 Bartolomeu Dias reached the Cape of Good Hope at the bottom of Africa; Vasco da Gama reached India on his voyage of 1497-99. This was perhaps the most significant of all the voyages of discovery: a seaward route to India was the beginning of colonialism, multiculturalism and globalisation. The world had new horizons: and all in order that the food on a rich man's table might taste a little richer.

There was also a theory that you could reach the spice-producing lands by sailing the wrong way: and Christopher Columbus, a Genoese sailing under the flag of Spain, set off westwards in 1492. He never reached Asia. He made his first landfall on the Bahamas, established a colony on an island he called Hispaniola, now Haiti, and also visited Cuba. Later voyages brought him to the mainland of South and Central America.

Ferdinand Magellan found his way round the great double continent of the Americas and passed through what are now the Straits of Magellan, between Chile and Tierra del Fuego. He went on to reach the Philippines, where he was

Botanical treasure: a cinnamon seller, from Tractatus de herbis, *France (fifteenth century).*

killed on the island of Mactan. There was a mutiny on the voyage, and much trouble followed the return of the survivors. It was not counted an instant success, but this was the first voyage to circumnavigate the earth – and perhaps more to the point, it established the western spice route. This was a trading route that worked effectively for more than two centuries and made Spain the dominant power in Europe. It finally became redundant with the opening of the Panama Canal in 1881.

Other voyages of discovery and then of colonisation followed, with France, Holland and Britain taking a major part. The rush was on: the rest of the world was there for European powers to exploit and the exploiters came in droves. The age of empire had started: and it all began with spice.

THIRTY-FIVE

KUDZU

I'm just a mean green mother from outer space and I'm bad.

Charles B. Griffith, *Little Shop of Horrors*

We humans have succeeded in domesticating around thirty species of animals: we control their breeding and choose what lives they should live. We have domesticated at least 1,000 species of plants. Some estimates put that number far higher: fully and semi-domesticated plant species have been calculated at up to 2,500.

The reason for this huge discrepancy is obvious enough: it's very hard to domesticate an animal, as already discussed in these pages. But you can domesticate a plant right now: just find a wild plant and take it home. Or seize the land it is growing on, if you are more ambitious. After that you grow it and manage its reproduction. Only propagate the plants you like: once you have imparted human choice onto the nature of a species, the process of domestication has begun. That is why domestic species in this book far outnumbers those I wrote about in *The History of the World in 100 Animals*: there are a very great deal more to choose from and most of the plants that have affected human history have been domesticated.

It can seem at times that we humans have established total control over the plant kingdom: a plant either grows because we choose to cultivate it or because we decline to destroy it. But there are still plants that defy our attempts to control the biosphere: awkward anomalies that ask questions not only about the extent of our abilities but about the extent of our wisdom. Kudzu is a classic example. It was first deliberately cultivated in the United States as a plant that would grow the nation out of Depression. Now it is known as the vine that ate the South.

James Dickey, author of the novel *Deliverance* and the script for the eponymous film and a former American poet laureate, wrote a poem about kudzu:

In Georgia, the legend says
That you must close your windows,
At night to keep it out of the house
The glass is tinged with green, even so

Yes, it's another plant that is coming to get you. Kudzu is a name given for a number of species of vine, all from the pea family: clambering plants with groping tendrils. They grow from a stout and complex root system that keeps throwing up new shoots. They can reproduce vegetatively (by cloning) as well as by producing seeds. Kudzu was brought into the United States for the Centennial Exhibition in Philadelphia in 1876, and it came to the southeast of the country at the New Orleans Exposition of 1884. It was marketed as the perfect plant for your porch, growing quickly to establish a pleasant green shade, with added useful properties as fodder for livestock and a preventer of erosion.

The plant is well set up for rapid growth, with much energy stored in its roots. It can notoriously grow 1 foot (30cm) in a day, and a stem can reach a length of 100 feet (30 metres). If you want a lot of growing plant in a hurry, kudzu is a good choice. It seemed the answer to the event known as the Dust Bowl. This was a combination of drought and inappropriate farming: deep ploughing on the former grasslands opened up the topsoil to the winds and wind erosion carried away the soil. It was an early example of an anthropogenic ecological disaster. Droughts took place in 1934, 1936 and 1939-40. One of the solutions put forward was kudzu. The plant grows fast and binds the soil: fixing it, and keeping it safe from the wind.

The cause of kudzu was pushed heartily by Channing Cope, a radio host and columnist for the *Atlanta Constitution* newspaper; he insisted repeatedly that the fields of America were 'waiting for the healing touch of the miracle vine'. Kudzu was planted with immense enthusiasm by the Soil Erosion Service and the Civilian Conservation Corps. Workers were paid to plant it: by 1946 a total of 3 million acres (1.2 million hectares) had been planted with kudzu.

That brings us to the law of unintended consequences: perhaps the most important factor in every human attempt to manipulate the environment. Industrialisation was intended to bring happiness and glory – or at least money – to human society; no one meant to start the process of climate change. And on a much smaller – but still considerable – scale, no one intended kudzu to be anything other than a good thing.

Kudzu is good at stopping erosion because it grows so fast. It follows that the plant is prone to growing beyond the borders established for it by humans. As the boll weevil devastated cotton farming (see Chapter 74) around the time of the Dust Bowl, many farmers abandoned their land in the Deep South. Kudzu moved in.

Kudzu is not a problem in Japan, where it is native; the name has been adapted from the Japanese *kuzu*. The plant grows on mountainsides and the annual growth is routinely bitten off by frost. But in the benign climate of the American South, kudzu is unstoppable. Around 7.4 million acres (3 million hectares) are covered by kudzu in the states of Alabama, Georgia, Tennessee, Florida, Mississippi and

Opposite Coming to get you: Vinelash by Mark Tedin.

North and South Carolina. It is a remarkable sight: buildings and telegraph poles draped in a cloak of green, forest edges blurred by a mountain of vine, which shades out the trees that support it and eventually causes their death. Kudzu is tolerant of stress and drought, and most estimates say it is increasing. United States Forest Services puts the annual increase at 2,500 acres (1,000 hectares), the Department of Agriculture at 150,000 acres (61,000 hectares).

It's very hard to kill. Repeated attacks on what you can see above the surface will eventually kill the roots, but it takes a good deal of time, whether you are working by close mowing or by heavy grazing with domestic animals. You need to remove the root crown to do the job at one go, taking with it all the rooting runners, but that's not easy to achieve. Herbicides are not an immediate and total success: soil-active herbicides, which are absorbed by the plant through the soils, are effective over ten years. Fungi have been introduced as a form of biological control. The most effective and least devastating method of control is grazing by goats, but this is not cheap or easy.

Kudzu has become something of a badge of the South: an emblem of perverse pride, one more thing that makes the South different. It remains a classic example of the way humans damage the environment we live in, often with the best possible intentions. The fields were intensively farmed to bring more food to humans and that created the Dust Bowl; the remedy for the Dust Bowl was arguably as bad as the problem. We create problems, we fix them. Then we have the problem of fixing the solution…

THIRTY-SIX

AMERICAN GRASS

The bastards have never been bombed like they're going to be
bombed this time.

Richard Nixon

When a tropical rainforest is felled for timber it can never be recreated as it was. As we have already noted, a tropical rainforest is not powered by the richness of the soil; it is driven by the richness of its own immense and biodiverse self. All the same, after the catastrophe of felling primary jungle, a second-rate forest can return in its place. Not everything is lost.

But if destruction rather than timber is the goal, total destruction is well within the human scope. Between 1961 and 1971, 5 million acres (2 million hectares) of rainforest and mangrove forest were destroyed in Vietnam, Cambodia and Laos. What remains is not nothing. It is worse. What remains is what the Vietnamese sardonically refer to as American grass: a thick tussocky growth that can be 10 foot (3 metres) high. It is more or less literally no good to man nor beast, and it's desperately hard to get rid of. It also chokes out anything else that tries to live.

The grass is mostly of two species, *Imperata cylindrica* and *Pennisetum polystachion*. They moved in after the destruction of the forests and established themselves in a series of impenetrable mats. You can't even walk through them. The forests that once stood there are casualties of war.

Trees are our enemy. Under that slogan the armed forces of the United States of America waged war on the forest itself. The Americans were involved in warfare in Vietnam for almost twenty years before pulling out in 1973. It was a war of bitter divisions even within the United States: in Vietnam the American forces were supporting South Vietnam against the North. The soldiers from North Vietnam operated as guerrillas, striking from cover and withdrawing swiftly. Often they hid in the forest: so the US forces operated on the principle that if you destroy the forest, you destroy the enemy. They called it Operation Ranch Hand. It was an idea borrowed from the British, who deployed it during the so-called Malayan Emergency, a civil war in what is now Malaysia between 1948 and 1960.

The Americans adopted a policy of ecocide in Vietnam and their chief weapons were defoliants. These were sprayed onto forest and agricultural land, mostly from the air. These were colour-coded and given the prefix Agent: thus Vietnam and its

neighbours were sprayed with Agents Pink, Green, Purple, White, Blue – and, in quantities far surpassing all these, Agent Orange.

The campaign against the forest used 21 million US gallons of herbicide (17.5 million UK gallons, 95 million litres), of which 60 per cent was Agent Orange. Of the land treated, 24 per cent got a double dose and 12 per cent triple. It was the largest chemical warfare project in history. It was arguably forbidden under the Geneva Protocol of 1925 (the Geneva Convention was established in 1949), which was an agreement not to use chemical weapons. So far as the Americans were concerned, Agent Orange was a benign tactical herbicide, harmless to humans and with short-lived effects on the environment.

Agent Orange contained dioxin, which is highly poisonous. It was present as a by-product in the process of manufacturing the herbicide; it is also produced by other processes, including the burning of rubbish, the burning of fossil fuels, smoking tobacco and bleaching. I must state here that this is a book about plants, and so this chapter's main focus must be the effect of Agent Orange on the plants it was aimed at. However, it would be irresponsible not to note in passing that the poisons dumped on the land with the intention of killing forests also affected the human population: dioxin causes cancer, birth defects, reproductive difficulties, problems in child development and with the immune system. It's been calculated that 2.1 million people were directly affected by dioxin; other estimates double that. Dioxin contaminated water and food and its effects are being felt into second and third generations. In the course of a decade, 880 pounds (400 kg) of dioxin were dropped onto Vietnam.

The effects of all this poison on the environment were massive and are still visible. Even now, fifty years on, it is clear that this is still a country in recovery. The destruction of the forests means that the ecosystem is far less robust, with an immense loss of biodiversity, which, as we have seen, is what gives resilience to any ecosystem. The soil has been impoverished, the water contaminated, without the trees there is a great deal more flooding and erosion, the nutrients in the soil have been leached away and invasive plants – most particularly American grass – have moved in. It looks hopeless. It isn't.

Defoliation was a campaign of breathtaking callousness, reckless of human lives, devastating in its extent and, what's more, it didn't even work. 'The Americans said they would bomb us back into the Stone Age,' said Pham Tuan Anh, CEO of the Vietnamese conservation organisation Viet Nature. 'They didn't.'

They didn't, and Viet Nature is involved in the regeneration of the forests. This is an extraordinarily difficult business. First the invading American grasses have to

Opposite The enemy: not the man but the plant, so-called American grass in Vietnam, photograph by David Bebber.

be removed: and because they grow in huge tussocks, it is impossible to use machinery on the areas affected. The task has to be done by hand, which is slow and desperately hard. In 2018 I had the privilege of visiting a site where the regeneration was continuing: the saplings that had been planted were now above head height and beginning to lose their look of vulnerability. The area had been planted: 1,650 trees per hectare, nine different species, 7 hectares in total. The mix included two fast-growing pioneer species, which fix the soil and provide protection for the slower-growing trees that will follow them up towards the sun.

This is a method road-tested in Thailand, where Viet Nature staff went to study it. It was already beginning to work. 'We're establishing a model here – so it's a very important project,' said Tuan Anh. 'We have the help and support of the local community and we have been able to help them.'

We took measurements. Impossibly, the American grass still surrounding the replanting project was 8 inches (20cm) taller than it had been three weeks earlier; a shame that the trees couldn't grow as fast. But as the new forest gains a foothold, so seeds from neighbouring forest will come in by the wind and in the droppings of birds. In this improving habitat, some of these will germinate. Here, in a landscape of ancient devastation, an area of hope had been established.

It was a powerful experience to be in such a place. It's about regeneration: the regeneration of a country after war, the regeneration of hope after despair. But regeneration is what life is all about: if you like, what life is for. The continuation of the world as a viable ecosystem depends on its ability to regenerate itself: or rather, on the extent to which the planet's dominant species allows it to regenerate. Here, in a classic example of deliberate destruction, we can find the beginnings of regeneration: also willed.

THIRTY-SEVEN

TOBACCO

All they that love not tobacco and boys are fools.

Christopher Marlowe

Plants make food. By doing so, plants allow us animals to exist, but they don't always make it easy for us. Across the millennia they have evolved and continue to evolve glorious and complex ways to avoid getting eaten. Animals, in their turn, have evolved and continue to evolve ways of combating their defences. It's an arms race that has been going on since the first act of consumption.

Plants fight back with spines, prickles, thorns, stingers, inaccessibility, timing, hard barriers like bark, hard shells, waxy coatings. Some employ insects to guard them, offering shelter in return. And many put out chemicals that are unpleasant, and even toxic. Some of these poisons are deadly. That should make them plants for humans to avoid at all costs, but sometimes the plants are useful to us. And sometimes the poison itself is deeply attractive to humanity. The nightshade family of 2,700 species includes potatoes (Chapter 32), tomatoes (Chapter 84), aubergines (eggplant) – and tobacco.

The tobacco plant protects itself with a substance called nicotine. When consumed, it affects the muscle response to signals from the brain; in sufficient doses it causes constant muscular contraction that leads to paralysis and death. Every living creature with muscles can be affected by nicotine. The tobacco hookworm, a caterpillar of a species of moth, has developed the ability to excrete the nicotine without being troubled by it. It also passes nicotine from the pores of its skin as a defensive mechanism for its own use, making itself unpalatable – a strategy that has been called defensive halitosis.

But most plant-eaters avoid the seventy-odd species of tobacco plants. Humans are not among them: across the millennia we have found their poisons deeply attractive. Tobacco originates in the Americas and was used in pre-Columbian times in religious ceremonies, as a painkiller, as a trade item and even as currency. It was brought back to Europe by the explorers, and is one of the four great New World crops that spread across the world, along with maize, potatoes and tomatoes. Tobacco is the only one routinely used in large quantities in every country on earth.

Tobacco plant: illustration by Giorgio Bonelli (b.1724) for Hortus Romanus, *published in Rome 1772-93.*

Tobacco was unknown in Europe before the first Columbus expedition of 1492. Tobacco smoking in Cuba was described by Bartolomé de Las Casas, a priest and historian who travelled there in 1502:

> Men with half-burned wood in their hands and certain herbs to take their smokes, which are some dry herbs put in a certain leaf, also dry, like those the boys make on the day of the Passover of the Holy Ghost; and having lighted one part of it, by the other they suck, absorb, or receive that smoke inside with the breath, by which they become benumbed and almost drunk, and so it is said they do not feel fatigue. These, muskets as we will call them, they call *tabacos*. I knew Spaniards on this island of Española who were accustomed to take it, and being reprimanded for it, by telling them it was a vice, they replied they were unable to cease using it. I do not know what relish or benefit they found in it.

Tobacco was brought to Europe, where it was praised as a great benefit to health. It was promoted by Jean Nicot, French ambassador to Lisbon, who took it to France and gave his name to its active ingredient. He claimed that tobacco cured thirty-six separate maladies. The consumption of tobacco was pleasant, calming, social and highly beneficial: who could fail to love it?

The principal property of tobacco, so far as humans are concerned, is not that it does you good, or even that it does you harm, but that it makes you want more. It is the perfect trading commodity: it offers nothing but the desire for further consumption. As its popularity spread, so the willingness to supply this new need increased.

Not everyone saw it that way. James I of England, a natural outlier, wrote *A Counterblaste to Tobacco* in 1604: 'Have you not reason then to bee ashamed, and to forbeare this filthie noveltie, so basely grounded, so foolishly received and so grossely mistaken in the right use thereof?' The king's wrath was not enough to affect the spread of tobacco consumption, in Britain or anywhere else, and so the newly formed American colonies grew more and more of it. The climate was perfect for the plant, but there was still a snag: the production of tobacco required a great deal of labour, with the cultivation, harvest and subsequent curing of the leaves. In the curing process the harvested tobacco was dried with straw and then hung on sticks or, later, taken into barns. It was a tricky business: the tobacco must stay moist, or it will crumble, but not too moist or it will rot. Once cured and dried it was transferred into barrels for transport.

Those that used indentured labour for all these processes found it hard to turn a profit. The answer was obvious: slavery. This was hardly a new thing in the New World, but the tobacco trade brought it onto the mainland of North America, particularly into Virginia and the Carolinas. In 1619, twenty African slaves were brought to Virginia: the impact of that first step is still being felt today. With slavery, tobacco farming became highly profitable; money from tobacco helped to finance the American Revolution and thus to found the United States of America.

Tobacco was of course smoked in pipes. It was also chewed, so that the spittoon became a standard piece of furniture. It could also be taken as snuff, and also as snus, a moist powder placed between lips and gum. The leaves could be rolled up and smoked as cigars, and the sweepings could be gathered up, placed on paper and rolled into a tube: a cigarette.

In 1881 James Bonsack invented a machine to do the rolling, and that made cigarettes the most convenient way of consuming tobacco. Smoking was first established as something men did. Women then became smokers too, and smoking became charming, elegant, agreeable, social. Films showed the most admirable people wreathed in smoke: Fred Astaire and Ginger Rogers with immense style, Humphrey Bogart and Lauren Bacall with indomitable toughness. Tobacco growing spread to Asia and Africa. In the First and Second World Wars, cigarettes were provided to the troops with rations. A friend of mine who did national service in the Royal Navy in the early 1950s remembers being issued with 600 cigarettes a month. There is a superstition, one dying out with the passing of matches, that to offer someone the third light from the same match is unlucky: it gave an enemy sniper an opportunity to aim and fire. The last words of the great Edwardian writer H. H. Munro, who wrote as Saki, were: 'Put that bloody cigarette out.' Lest I sound a little sanctimonious here, I should add that I smoked Gitanes and Gauloises with great enthusiasm in my early twenties.

The first people to take action against tobacco were the Nazis, who condemned the practice, taxed it heavily and banned smoking in public. The first studies into the possibilities of health damage from cigarettes began in 1948 under the British physician Richard Doll; a connection between smoking and lung cancer was established in 1950. In 1964 the surgeon general of the United States, Luther Terry, published a report entitled 'Smoking and Health'. Health warnings on cigarette packets and increasing restrictions on advertising and later sponsorship were introduced in many countries. The tobacco companies responded by targeting women: the brand Virginia Slims was especially created for women; the company sponsored the nascent women's tennis tour and so played a small but significant part in the rise of feminism. Cigarettes were increasingly advertised as low tar, or in some other ways less damaging. Kent cigarettes laid great

Take the air in a tobacco trance: woman smoking a Havana cigar.

stress on their micronite filter, which contained asbestos. An advertisement on United States television claimed that 'more doctors smoke Camel than any other cigarettes'; it showed a man in a white coat enjoying a cigarette between patients.

The tobacco companies also increased their marketing push in developing countries, linking cigarette smoking with status, achievement and masculinity. I remember a television advertisement for Viceroy Super Longs from the late 1970s when I lived in Hong Kong; it showed a white-suited Chinese man leaning on the taffrail of a yacht, a blonde lady in the background: 'Everything I have I've earned. Success in America. The yacht. The right to smoke the cigarette I choose…'

Tobacco is the world's most smuggled legal product. The World Health Organization says that tobacco consumption is the world's single largest cause of avoidable death. Tobacco smoking is associated with diseases of the heart and lungs, heart attack, stroke, emphysema, cancer of the lung, mouth and pancreas. The dangers of passive smoking were established in the 1990s. The addictive potential of nicotine is reckoned to be almost as high as that of the opioids (see Chapter 57). It's been calculated that 80 per cent of the word's 1.3 billion smokers come from low- and middle-income families.

Nicotine was developed by plants to poison insects. This function has been exploited by humans for the same purpose with the development of neonicotinoid insecticides, which are derived from nicotine and are structurally similar. The poisons were supposed to be safe and ecologically friendly, but they kill insects efficiently and indiscriminately, causing populations of insects to crash, and with them the birds and other animals that consume them, causing immense damage to the ecosystems in which they are used. The three main types of neonicotinoids were banned in the European Union from 2018. In 2014 the United States under Barack Obama banned the use of neonicotinoids in Wildlife Refuges; a decision reversed five years later under Donald Trump. In the UK, the ban on neonicotinoids was lifted and sugar beet seeds can now be dressed with the stuff.

Smoking has become less acceptable in the developed world, banned in most public places, including restaurants and bars. People can smoke outside pubs in Britain, meaning that you need to go inside to get a breath of fresh air. The pariah status of the smoker has created a perverse pleasure for some. Some people smoke because tobacco suppresses the appetite and therefore helps you to stay thin. In the main, smoking is on the decline among the well-off: but the tobacco industry is not in decline globally. It has merely shifted its ground. China gets up to 10 per cent of its revenue from the China National Tobacco Company, which has a state monopoly and supplies 350 million Chinese smokers. Tobacco is the most marvellous of all products for trade: as Oscar Wilde said: 'You must have a cigarette. A cigarette is the perfect type of a perfect pleasure. It is exquisite, and it leaves you unsatisfied.' But you can always have another.

THIRTY-EIGHT
LOTUS

No mud, no lotus.

Thich Nhat Hanh

The idea that beauty is in the eye of the beholder is proverbial: we are eager to accept personal choice on what pleases. A hundred people walking round the National Gallery in London will come away with getting on for 100 personal favourites. But some kinds of beauty seem more or less universal: as if there was an innate and collective – rather than cultural and personal – idea of beauty. It brings with it a powerful tendency to equate that beauty with truth.

We mostly find this consensus of beauty in the natural world: a rose, as we have seen in Chapter 3, a bird of paradise or a blue morpho butterfly. We also find it in the flower of a lotus. Few humans can gaze on a fully open living lotus flower without feeling that this is somehow a good thing.

A lotus doesn't just happen to be attractive. It is attractive in a purposeful way: attractiveness is its evolutionary purpose. It is there to attract insects, because without insects it can't make more lotuses. In the same way, the colour of a butterfly's wing and the plumage of a bird of paradise are *supposed* to be attractive: they exist so that the possessor of this beauty will be able to propagate itself. A root is not attractive; it is not supposed to be and it has no need to be. Few people see beauty in a root, even though it is packed with meaning. But a flower, though its purpose is to please non-human eyes, gives delight to humanity. We gaze on its beauty and feel that truth must lurk there.

The lotus is sacred in Hinduism, Buddhism, Sikhism and Jainism. Two species are recognised. Neither is closely related to waterlilies, though they have a similar way of life; it's an example of convergent evolution. Lotuses live on floodplains and in slow-moving rivers and deltas in central and northern India up to 4,300 feet (1,400 metres) in the Himalaya, in Sri Lanka and Southeast Asia; they have spread, probably by human hands, into New Guinea and northern Australia. They have flowers up to a foot (30cm) across, and the plant can warm its flowers up under its own power, like a warm-blooded vertebrate. A lotus has been recorded

Opposite Sacred plant: a woman seated on a lotus flower, visual depiction of Indian music, nineteenth century (School of Rajasthan, artist unknown).

at 35°C (95°F) when the ambient temperature was 10°C (50°F). This warmth is attractive to the insects in itself; it also serves to push out the scent of the flower and advertise the plant's desirability. The heat also energises the (cold-blooded) insects for further flights, collecting and delivering pollen.

Lotuses live in water to a depth of up to 8 feet (2.5 metres). Such waters can come and go, with dry seasons and changing river courses; the seeds a lotus drops remain viable for an exceptionally long time; the record is 1,300 years. Lotuses can live in almost anoxic water – water containing no dissolved oxygen. They compensate with a system of internal gas canals, in which oxygen is taken down from the air by older leaves on the surface of the water, to be transported down to the rhizomes in the mud below. This ventilation system passes carbon dioxide out by the same system, and it exits through holes called stomata in the youngest leaves.

Lotuses are cultivated for food in Asia; their rhizomes, seeds and stems are eaten. In *The Odyssey*, Odysseus lands on the island of the Lotus-eaters, where those that eat this food are filled with a blissful forgetfulness. The affected sailors had to be dragged back to the ships and chained up; it's been suggested that this is more likely to refer to opium poppies (Chapter 57) than the lotus that we know.

Many depictions of Hindu gods and goddesses show them sitting on a lotus throne; the lotus stands for perfection and, if you like, for what is immortal in humanity. Vishnu, the preserver and guardian of humans, is often referred to as 'the lotus-eyed one'; Brahma, the creator, was born from a lotus that emerged from Vishnu's navel. It's said that when Prince Gautama, later the Buddha, walked on earth, lotuses appeared in his footmarks. But when he was asked if he was a god, he compared himself to a lotus: 'Just like a red, blue or white lotus – born in the water, grown in the water, rising up above the water – stands unsmeared by water, in the same way I – born in the world, grown in the world, having overcome the world – live unsmeared by the world. Remember me, Brahman, as awakened.'

The leaves of the lotus rest on the surface of the water; the area of plant available to carry out the task of photosynthesis is directly related to the area of the water it can cover with leaves. The plant doesn't reach out and branch in the manner of a tree; it is dependent on water to support its weight. The leaves are on long, flexible stems without rigidity, so even if the depth of the water varies, the leaves are always floating on the surface, their faces turned to the sun. It's an economical system that gets the best from its watery environment. But the plant needs to be pollinated by flying insects, and so the flowers are equipped with rigid stems. These take the flower above the water, which can be silty, anoxic and foul-smelling. It's above all this that the lotus rises: more eloquent to human minds even than the beautiful rose standing in manure.

The idea of rising above the troubles and the attachments (important Buddhist word) of the world is central to Buddhism of all kinds, and the lotus, with its utterly straightforward beauty, is a symbol that has lasted across the millennia. A Zen saying has it: 'May we exist in muddy water with purity, like a lotus.' The origin myth of Zen Buddhism concerns a silent sermon preached by the Buddha. In this, the Flower Sermon, he held up a single flower, naturally (in most versions) a lotus. Most were puzzled by this, but one disciple, Mahākāśyapa, smiled in recognition of the truths revealed.

Enlightenment, purity, rebirth and triumph over obstacles, expansion of the soul: all things can be seen in the flower of a lotus. The best way to meditate, it is said, is to assume the lotus position, seated cross-legged with the soles of the feet upward, heavenward. The Sanskrit word for lotus is *padma*; the one Sanskrit mantra widely familiar in the West is *om mani padme hum*, which means, roughly, 'the jewel in the heart of the lotus'; the chiming 'M' sounds are an aid to deep meditation. One of the most important Buddhist scriptures, according to some traditions the final teachings of the Buddha, is the Lotus Sutra, which is a practical guide to almost everything and includes the notion that every living thing can achieve enlightenment.

John Keats completed his 'Ode on a Grecian Urn' with the lines:

Beauty is truth; truth beauty – that is all
Ye know on earth, and all ye need to know.

It is clear enough that beauty matters to us humans, and no doubt to many other species, on a relatively deep level. We see a lotus flower, and whether or not we take in the fact that it is rising above noxious waters, we cannot help but be struck by its beauty. We don't see it just as an indicator of tasty rhizomes; we don't just see it as the sex organs of a plant. We see it as something that matters for itself: for a beauty that transcends function. We want, perhaps we need, beauty to show us truth, to help us to understand life better and live our lives better. It is in our human or our animal nature to seek beauty. Is the quest for beauty some kind of survival mechanism? Certainly, life is not much without it.

THIRTY-NINE

MARIGOLD

Open afresh your round of starry folds, ye ardent marigolds!

John Keats

About 100 million years ago there was a great explosion of plant diversity. This was the time that flowering plants really took hold. It was a radically new strategy and it changed the way the earth looks. Today there are about 350,000 species of flowering plants or angiosperms; that is to say, plants that wrap up their seeds in a fruit. (Gymnosperms, which include conifers, have unclad seeds; a gymnasium was a place where you were naked.)

Before this explosion took place, the job of exchanging genetic material was mostly done by the wind. Many species of angiosperms are wind-pollinated to this day, notably the grasses, including the plants we grow as cereal crops. The system has its drawbacks: it's dependent on chance, and it lacks a precise policy on targeting. To develop a target policy you need targets, as we saw in Chapter 13 on orchids – and that is exactly what flowers are.

Plants developed coloured leaves to stand out from the sea of photosynthesising green: these leaves sent out an unambiguous message: here I am, come and get it. Insects and other animals could now be directed onto the sexual organs of plants. They were (mostly) paid in nectar, drank their fill and departed, carrying pollen with them. They then moved off in search of another flower and there, if all went well, they deposited pollen and so allowed the plant to make more of its kind. The extravagance of flowers is actually a kind of economy: it is cheaper to provide a few gorgeous blooms in the high probability that the pollen they contain will reach its target, than to produce massive amounts of pollen and hope for the best. In grasslands, where plants of the same kind stand in close proximity, the system of wind pollination works well enough; when the plants are more scattered, it is an infinitely better strategy to employ an agent to do the work with greater precision than the wind could ever manage.

And, as we have seen with the rose (Chapter 3) and the lotus in the previous chapter, many flowers are deeply attractive to humans, even though we (now) have little interest in their nectar or their pollen. In many different cultures people use flowers to mark the most important things in life. We use flowers at births, weddings and funerals; we use them for religious celebrations, we use them as

Flower of the dead: The Offering, *1913, by Saturnino Herrán (1887-1918).*

love-gifts, for apologies, for thanks. The annual retail value of the cut and pot flower industry in the UK is £2.2 billion.

We express this passion for flowers in many different ways. In Mexico, the *Dia de los Muertos*, the Day of the Dead, is celebrated with marigolds. Exotic details like sugar skulls fascinate people from other cultures, but this annual celebration of the beloved dead is mostly conducted by means of millions upon millions of marigolds. Marigolds look like the sun: they seem to bring the warmth of the sun with them, and their flowers light up the vigil in honour of the dead. They bring life, for this is a time of celebration, not mourning: it is about the continuation rather than the cessation of life. The indomitable brightness of the marigolds stresses that this is a time of jollity: a celebration of the people you loved and continue to love. So you drink the favourite drink of your dearest departed, and against the background of marigolds you celebrate: for who can be morose in a room full of flowers bright enough to hurt your eyes?

In English they are called marigolds because they're Mary's gold, the treasure of the blessed virgin... but we need a spot of clarification. There are a good few species bearing the name of marigold. Here we will mostly be considering Mexican species from the genus *Tagetes* (these include those known as Aztec or African marigold, and another known as French marigold) and a North African and southern European species *Calendula officinalis*, sometime called pot marigold, otherwise ruddles, common or Scotch marigold; this was cultivated for its beauty and it reached England perhaps 1,000 years ago. Both these genera are from the family *Asteraceae*, which includes sunflowers (Chapter 8).

There is no one story for the flower's association with Mary, other than its striking colour and obvious beauty. The rays of golden petals mirror the rays of light coming from the head of Mary in many representations of her. There is a story that when she was fleeing the wrath of King Herod with Joseph and the infant Jesus, she was set upon and her purse was stolen. When the robbers opened it, it was full of marigolds. People made gifts of marigolds – in lieu of money – to altars and shrines sacred to Mary. The flower opens and closes daily, in the manner of the daisy (Chapter 10). In closing it often traps a little dew, which it releases on opening, so that you might imagine that the flower was weeping. In *The Winter's Tale*, Perdita talks of

> *The marigold that goes to bed with the sun*
> *And with him rises weeping...*

... which is a salutary lesson for all virgins.

The *Tagetes* marigolds made it to India, where they now play a central part in religious festivals and in the sanctification of daily life. They got there after the Europeans first reached the Americas, and received much of the New World's bounty in the process known as the Columbian exchange; India got chillies (Chapter 40), potatoes (Chapter 32), tomatoes (Chapter 84) and marigolds; wheat (Chapter 2) and apples (Chapter 24) went the other way. Marigolds have only been in India for about 350 years, but they feel like an ancient part of daily life. You can buy a garland of threaded marigold flowers any morning you choose to walk through the market: you can wear them, hang them on a shrine to Ganesh, even as he sits on his lotus throne, and you can decorate the windscreen of your lorry with them. Marigolds add gaiety and a whiff of prayer to everything they touch. The 2011 film, *The Best Exotic Marigold Hotel*, was about Anglo-Indian cultural clashes; the choice of the title was perhaps inevitable.

Marigolds produce chemicals that repel certain damaging insects, notably aphids like greenfly and blackfly, and also deter damaging nematode worms around the roots – and that is another good reason for having them around. The Mexican custom of arraying corpses with marigold flowers has more than aesthetic value. *Tagetes* marigolds are often used as companion plants in vegetable gardens, keeping damaging insects away without need for commercial insecticide. The *Calendula* species has been used to treat eye complaints related to age: cataracts and macular degeneration. They are an ingredient in ointments for the skin and have been used in cooking, adding flavour to wine and porridge, a substitute for the more expensive saffron (Chapter 68). *Calendula* can also be used to reveal fairies: all you have to do is burn the plant in the form of an incense.

FORTY

CHILLI

Love is a burnin' thing
And it makes a fiery ring…

'Ring of Fire' by June Carter Cash and Merle Kilgore,
sung by Johnny Cash

Christopher Columbus believed that he had sailed to Asia and found a new source of pepper. In fact, he had reached the Americas and found the quite unrelated chillies. Nonetheless, the islands were called the West Indies and the plants peppers.

Chillies – or chilli peppers – were new to Europeans, but they had been cultivated in the Americas for at least 8,000 years, and no doubt consumed regularly long before that. They are, with potatoes, tomatoes and tobacco, yet another member of the nightshade family. Their interest for humans is in their fruit – technically berries, because they are fleshy, lack a stone or pit and are produced by a single flower. These berries contain capsaicin, which, along with related substances (capsaicinoids), makes them taste hot.

There are two possible evolutionary reasons for this. The first is that capsaicin kills fungus, and is therefore beneficial to the plant, because it repels attacking fungi. The second is that it puts mammals off eating the fruit. Capsaicin binds with pain receptors in the mouths and throats of mammals, causing them to send a message to the brain: this food is dangerous, don't eat it. It has no effect on birds, who lack the pain receptors. Birds are better than mammals at distributing the seeds, because they swallow them whole rather than chew them up. Seeds passed out by birds are more likely to be viable than those excreted by mammals. No doubt both aspects are helpful to the plant; you can flip a coin to decide which prompted the evolution of capsaicin.

This formidable defence mechanism turned out to be highly attractive to humans. The sensation of burning releases endorphins – the body's soothing response to pain. The same thing happens to people who exercise hard. Eating chillies is a way of experiencing runner's high without the actual running. Both experiences have an addictive side to them.

Opposite Fire starters: chillies from Ernst Benary's Album Benary, 1876-82.

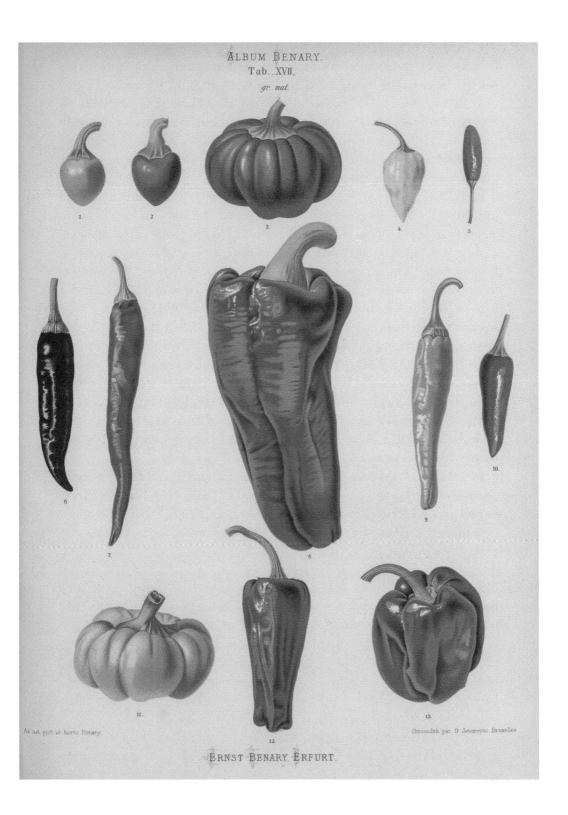

ALBUM BENARY.
Tab. XVII.
gr. nat.

1. 2. 3. 4. 5.
6. 7. 8. 9. 10.
11. 12. 13.

Ad nat. pict. in horto Benary.

Chromolith. par G. Severeyns. Bruxelles.

ERNST BENARY ERFURT.

169

Columbus brought chillies to Europe, and they fed the craving for spices (see Chapter 34 on cinnamon). Portuguese mariners used them as a trade item; they reached India via the Portuguese enclave of Goa, which the Portuguese claimed in 1510. Chillies, essential to Indian cuisine, have only been known there for 500 years, but they fitted in very happily with the much older tradition of well-spiced (masala) food, one that creates a spectrum of spice flavours rather than a single blast of heat. Chillies moved on into much of Asia, into China through Macao, another Portuguese possession.

Chillies are a necessity in some culinary traditions, in many others a standard ingredient. You need them for arrabbiata and puttanesca sauces in Italy, as well as 'nduja sausage, for patatas bravas and chorizo sausage in Spain, in the mild form of paprika in much of Hungarian food. Chillies are essential to jerk chicken in Jamaica, many Sichuan dishes, and much of the cuisine of Thailand, Malaysia and Indonesia. And of course, in Mexico and India, the chilli is ubiquitous. Chillies are found in the harissa paste of North Africa, and in many African dishes further south. It's likely that chillies reached what is now the United States and Canada not by rising north from Mexico, but by travelling from Africa on the slave ships.

Five species from the *Capsicum* genus are used widely in cooking, and three of them dominate, coming in many different varieties and different degrees of heat. *Capsicum annuum* gives us bell peppers, which aren't hot at all, along with cayenne and jalapenos, which are; *C. frutescens* gives us tabasco and piri-piri, while *C. chinense* gives us most of the very hot varieties: naga, habanero and Scotch bonnet. There are at least 400 varieties in common use.

In 1912 an Americana pharmacist, Wilbur Scoville, invented the Scoville Organoleptic Test to measure the heat in chillies. The pungency of chillies could now be read in Scoville Heat Units, and different varieties of chillies could be given an objective rating, taking some of the Russian Roulette element out of cooking with chillies. Some examples:

	SHU		SHU
Bell pepper	0	Thai pepper	100,000
Pimiento	500	Scotch bonnet	350,000
Poblano	2,000	Trinidad Moruga Scorpion	1,000,000
Cherry Bomb	5,000	Carolina Reaper	2,200,000
Chipotle	10,000	Pepper X	3,180,000
Ring o' Fire	20,000	Pure capsaicin	16,000,000
Large red thick cayenne	50,000		

Chillies have been used to create pepper spray of up to 3,000,000 SHU, which is used in policing, riot control and self-defence. It is illegal for public sale in many countries. It was notoriously used by police in the United States on peaceful protesters at the University of California, Davis in 2011. It was also used in 2020 to disperse Black Lives Matter protestors, and again to clear protestors for a public appearance by President Donald Trump at St John's Church in Lafayette Square, Washington, the occasion he posed for cameras with a Bible. The spray was originally developed for defence against bears; many people who work in bear country carry pepper spray.

There is a fierce competition among chilli growers to come up with ever-hotter chillies, items which go way beyond the pleasure principle when it comes to consumption. There is a tradition of colourful names: Devil's Tongue, Dragon's Breath and Satan's Kiss. The top three – though there is much argument – are Carolina Reaper, Trinidad Moruga Scorpion and Pepper X. There are also extreme chilli sauces including Hell Unleashed and Chilli Hills Hell Hot Hand, which comes in a box shaped like a coffin. The experience they offer is more like bungee jumping than gastronomy: seeking terror with little risk of death and damage. It's a nice paradox that chillies are also used in pain relief: creams containing capsaicin give help to people suffering from the pain of arthritis and shingles.

Chillies are increasingly useful in the reduction of clashes between people and elephants in Africa and elsewhere. Elephants hate chillies: fences laced with chilli-soaked string have been effective. A buffer-zone of chilli bushes around other crops keeps elephants away. Elephants can also be shot at with paint-ball guns, using projectiles filled with chilli juice. I am a trustee of Conservation South Luangwa, based in Zambia, which, among many other initiatives, encourages farmers to grow chillies as a cash crop.

Let us conclude with a recipe for chilli sauce. Put, say, eight decent-sized tomatoes and half a dozen chillies into a whizzer. Choose chillies that meet your own tastes on the Scoville scale, but you need more than you'd expect. Add a few cloves of garlic, salt and honey (or sugar), and then whiz. Put some olive oil into a saucepan, heat it up, then add your sauce. Cook it for ten minutes or so and then let it cool. It'll keep in the fridge for a couple of weeks; it'll make the food taste good – and it'll keep most of the elephants away.

FORTY-ONE
TRUFFLE

If I can't have too many truffles I'll do without truffles.

Colette

James Bond, dyspeptic and out of sorts, muses grumpily on the 'French belly religion' in *On Her Majesty's Secret Service*. The author of this religion's sacred text, *Physiologie du gout*, is the nineteenth-century lawyer Jean Anthelme Brillat-Savarin, who said: 'The truffle is the very diamond of gastronomy.' Bond reaches for the Bisodol.

Truffles are fungi from the genus *Tuber*, and are, weight for weight – gram for gram rather than pound for pound, unless you're a millionaire – the most expensive food in the world. There is a 1999 record of a single truffle selling for US$330,000, a lot of money for a mushroom. Such prices are not the whole story: you can buy a 50g truffle for around £55, enough for a starter for four people, or a fairly lavish main meal for two. Even at the lower end, this is not stuff to chuck around.

When non-scientists talk about fungi, we mostly refer to the parts we can see: the fruiting bodies that we call mushrooms and toadstools (Chapters 51 and 61). The real existence of the fungus is in miles of tiny threads called mycelia that exist underground and inside organic material like tree trunks. Truffles are fruiting bodies of *Tuber* fungi.

Truffles are different from more familiar edible fungi because their fruiting bodies exist entirely underground. This might seem odd, since we expect fungi to send their spores off through the air, often in dense clouds; the giant puffball is named for exactly this trait. But truffles exploit members of the animal kingdom for dispersal. Truffles are *meant* to be dug up and eaten; they will be carried an appreciable distance from the site of the parent fungus within the consuming animal and when excreted, their spores are still viable.

Pigs, sows in particular, are notoriously efficient at locating these underground fungi. That's because truffles contain androstenone, which is also found in the testes and the saliva of boars. (Wild swine – usually referred to as wild boars – are the original target species of truffles; domestic pigs all trace their ancestry back to wild swine.) This substance has been found to affect humans too: in an experiment,

Opposite *Buried treasure: truffle-hunting shown on an educational card, late nineteenth or early twentieth century.*

HISTOIRE ANECDOTIQUE DE L'ALIMENTATION

LES TRUFFES

male and female subjects were shown pictures of normally clad women and asked to rate their attractiveness. The group that had been exposed to the odour of androstenone consistently rated the images higher than the control group did.

Pigs find truffles; the problem for humans is that they then dig them up and eat them – and pigs are hard to restrain. Their digging can disrupt the system of underground mycelia and make the truffle-grounds less productive. That's why dogs are trained to hunt for them instead: they can be dissuaded from digging and rewarded in ways that don't involve the consumption of fungus.

Truffles have always been associated with aphrodisiacal qualities. In Ancient Egypt there was a story of a childless farmer who observed a pig eating a truffle. When the pig didn't die, he tried one himself; he had thirteen children before he died. Truffles were prized in Greece and Ancient Rome. Pliny the Elder considered them miraculous, because they spring up without roots; they were praised by both Plutarch and Juvenal, who associated them with thunder and lightning.

Truffles then went out of favour for many centuries. They were associated with witches and frowned on by the church. But as the church's power first began to wane during the Renaissance, truffles became popular again; Catherine de' Medici and Lucrezia Borgia both served truffles at banquets. They were highly prized by Louis XIV of France, who attempted to cultivate them but without luck.

It's very hard to cultivate truffles. They are slow-growing and their mycelium needs to live among the roots of trees – oak, beech, birch, hazel, hornbeam, poplar. It can take ten years before the system is productive. Brillat-Savarin said: 'The most learned men have sought to ascertain the secret, and fancied they discovered the seed. Their promises, however, were in vain, and no planting was ever followed by a harvest. This perhaps is all right, for as one of the great values of truffles is their dearness. Perhaps they would be less highly esteemed if they were cheaper.'

Truffles were eventually produced commercially in greater quantities and were very popular in the nineteenth century. They were described for science for the first time in 1831; apparently Linnaeus was not a truffle-eater. But in the twentieth century, with the devastations of war, the increasing industrialisation of Europe and the consequent flight from countryside to city, forests were greatly reduced and it was much harder to find truffles. As a result, they went up in price and desirability. These days they are grown commercially in a number of countries, including Australia, New Zealand and the United States.

They are still not produced in quantities; there is no danger of them becoming democratic. They retain a mystique. Brillat-Savarin said: 'The truffle is not a positive aphrodisiac, but it can upon occasion make women more tender and men more likeable.' Sensuous, naughty and above all expensive, truffles are the epitome of belly-religion, no matter which country it is practised in. You are recommended to eat them with the plainest of plain food, to allow the flavour of the truffle to dominate proceedings: with pasta, rice, eggs or potatoes. However, if you wish to enjoy the flavour of wealth rather than mere fungi, you can insert truffle shards beneath the skin of chicken, turkeys, duck and guinea fowl. Best to acquire a truffle-shaver before you start cooking.

I have enjoyed truffles when dining with colleagues on the road, on those rather grim all-male occasions when every diner feels the economic pressure to eat and drink his fair share of the bill. A dish of truffled pasta offered a rare occasion when the vegetarian option outpriced the meat on the menu. I have found the dish very pleasant, though inferior in flavour to *funghi porcini*. But the price gives it a zing no surface-dwelling mushroom can rival.

FORTY-TWO

MARULA

*'Believe me, my young friend, there is nothing –
absolutely nothing – half so much worth doing as simply
messing about in boats.'*

Kenneth Grahame, *The Wind in the Willows*

The first humans walked the savannahs of Africa 2.5 million years ago. It wasn't until 100,000 years ago that they started to leave. The diaspora is one of the great events of human history: the beginning of the changing of the world. Many of those who left travelled by water. By canoe.

Wherever you find humans and water you find boats. Once you're on the water you can find more fish – and perhaps even more importantly, you can get to the other side of the river or further along the coast. You sit on a fallen tree and punt it along with a branch. When the water gets too deep you can use the branch to push against the water... and you have a boat. Everything that follows is a refinement of that idea.

The oldest known boat is the Pesse canoe, which was found in Holland and is about 10,000 years old. But humans were canoeing many millennia before that: the problem is that wooden artefacts don't preserve well. We call it the Stone Age because people used stone tools, but tools made from less durable substances rotted themselves out of recorded history (see rope in Chapter 12 on cannabis).

Evidence for canoes has been found in Australia 40,000 years old, and in Flores (part of Indonesia, the Lesser Sunda Islands) 90,000 years old. The fact is that canoes must have existed in prehistoric times. Once you have used a log it is no great leap to modify it. Once you have the tools you can change the world.

You can strip the branches from your fallen tree to make it move more smoothly on the water. You can fashion paddles that work better than simple branches. And then you can hollow out your tree trunk and make a canoe. This is a long and difficult job, especially if you are restricted to tools of bone and stone, but at least you have control of the process. You can select the right tree; and if you have tools you're no longer restricted to trees that have fallen by chance. Naturally, you choose the tree with the best wood for the job: a compromise between lightness and robustness. In Africa you could choose a hardwood like mopane; it would last forever but mopane is hard to work and doesn't float well. Other trees – baobab for

Portal of discovery: Sable Antelopes and Marula Tree *by Barbara Phillips.*

example – are light and float like a dream, but the wood is too soft and fibrous to be durable.

In many parts of Africa dugout canoes are still made and used on a daily basis. Three of the best trees are marula, kigelia (see Chapter 22 about making fire) and wild mango. I have chosen the marula to represent this trinity; to represent the principle of using plants for water transport. I have also chosen it because an old bush legend states that elephants routinely get drunk on fermenting marula fruits.

The marula tree is deciduous; it sheds its leaves and shuts down in the dry season. That makes the initial work on the canoe easier. You strip the branches and shape the trunk into an appropriate hydrodynamic curve, a process that requires a good eye and a steady hand. Even today, this complex task is often performed with homemade tools; the modern African is as brilliant at improvisation as the most ancient members of our species. They often use tools made from scrap metal and other useful items; nothing gets thrown away here. The best tool is an adze: an axe with the blade set at right-angles to the handle. The really difficult task comes with the hollowing out. A perfectly made canoe has an even thickness throughout: this is crucial to the balance and the performance of the boat. You can help the waterproofing by applying beeswax. You can also use fire, which hardens the sap into a substance like resin and again waterproofs the boat. This process makes the canoe more durable – and also more resistant to salt water. Don't let the canoe dry too quickly during construction or the wood may distort and crack. Better to wet the canoe again and again, leaving it to dry repeatedly until it becomes watertight. It takes a single person about six weeks to make a decent canoe.

The basic canoe is a tree – tree with style. The use of such craft opened up new food sources, made it possible to travel and to trade over greater distances and added greatly to the possibilities of sociability, inter-marriage and warfare. Canoes allowed different human communities to meet: and thus to increase cultural and genetic diversity. Over many thousands of years, canoes allowed our species to shift from local to global.

With increasing sophistication boats were made of other materials, including reeds and tree bark. People added sails, which drastically increased the boat's range, and outriggers, which increased stability, especially useful at sea. War canoes have been made up to 130 feet (40 metres) long; in 1506 the Moravian printer Valentim Fernandes reports canoes 80 feet (24 metres) long transporting entire communities, with two or three cooking fires going as the people moved on across the water.

The use of trees for travel allowed the civilisation of the Indus Valley to connect with that of Mesopotamia. They allowed the people of South America to reach the islands of the Caribbean. Canoes are especially associated with the Native Americans of what is now Canada and the United States; a canoe is sometimes termed a Canadian, or Canadian canoe, to distinguish it from a kayak. Kayaks were invented by the Inuit people, who made them from animal skins stretched over a framework of wood or whalebone.

The marula tree is also prized for its fruit. The kernels contain seeds which also make good eating; these are traditionally obtained by using two stones as hammer and anvil. They open more easily after they have passed through an elephant, so foraging elephant dung for the kernel's seeds is a profitable exercise. Elephant dung is the least disgusting of all: the food they take in is very lightly digested – that's because they are non-ruminant plant-eaters. The fruit is used to make a liqueur, Amarula, which contains cream and sugar. Alas, elephants are too big and the supply of fruit too small for the tale of drunken elephants to stand up.

I have walked the savannahs of Africa and crossed rivers on dugout canoes: an adventure for me that was once the daily routine of our species, and for many people still is. These days, in the developed world, canoes (and kayaks) are mostly made from plastic or fibreglass and used for recreation. I have such a craft myself: in the twenty-first century canoeing remains an atavistic pleasure.

FORTY-THREE
CHERRY

Ours is a world of suffering
Even if cherry-flowers bloom

Issa

Beauty drives us. It is as important as sex, food, home, offspring: and is perhaps inseparable from them all. We seek beauty because beauty is life: it makes life bearable; it makes life possible. What Dr Johnson said of a good book is true of beauty: it teaches us how better to enjoy life and how better to endure it.

We cultivate many plants for their beauty rather than for their material virtues. The marula tree (see previous chapter) gives us transport and sweet fruit: the cherry gives us sweet fruit and a fine red wood. But we have gone to immense trouble to create varieties of cherry that bear no edible fruit. We are satisfied with their beauty: they give us nothing more but they are cultivated, cherished and reverenced.

Many wild cherry trees – there are several species native to North America, Europe and West Asia – produce a nice soft fruit with a stone (pit), technically a drupe. The tree comes into flower before the leaves arrive, offering a beeline directly to nectar and pollen without distracting leaves. This strategy can only work with a brief flowering season: a tree can't lose too much opportunity for photosynthesis. The flowers come in small corymbs: that is to say, the stalks of the outer flowers on each cluster are increasingly longer than those inside. No flower hides behind another: they are all equally offered to the pollinators – and it's one of the reasons cherry blossoms are considered particularly beautiful.

Most of the fruit-producing cherries that we cultivate come from a single species, *Prunus avium*. This species probably made its way to Europe from Turkey. They are Goldilocks plants: they do best in winters that are not too cold but not too hot either; they need a little chill if they are to bloom the following spring. They also need moderately cool summers: and when all this is just right, they thrive. The best cherries I ever ate were a white variety in Armenia, served by the local mayor with Mount Ararat brandy.

Cherries have long been cultivated for their flavour and the sensuous pleasure of their consumption. Pliny the Elder mentions them, noting that Lucius Lucullus, politician, general and great promoter of learning, brought cherry trees to Rome

Vision of beauty: blossom viewing from Chiyoda Castle (Album of Women), *1895,*
by Yōshū Chikanobu (1838-1912).

in 74 BC. 'The cherry is among the first fruit to pay its yearly thanks to the farmer. It likes a north-facing position and cold conditions.'

They are a relatively expensive fruit in modern times, because they need a fair amount of looking after: irrigation and spraying as well as the labour of harvesting. It's a difficult fruit to grow well; the trees are prone to attack by aphids (cherry blackfly), fruit flies and fungi. Their faint sourness gives them a good depth of flavour in cooking: they are twinned with almonds in traditional dishes like Bakewell tart.

The tradition of growing cherry trees for their blossom alone began in Japan. Neither the wild species that grow there (*Prunus serrulata* and *P. speciosa*) nor their cultivars produce a fruit that humans find worth eating. There is beauty enough even in the idea: you devote a great deal of space and care to cultivate a tree that will yield nothing for human use but its beauty: and that for no more than a week. It is a reminder, delivered with a reasonably light touch, of the fleeting nature of both beauty and life: of happiness and sadness both. All this fits in with interlaced Buddhist and Shinto traditions. The notion of *sakura*, the cherry blossoms, goes impossibly deep in Japanese culture.

This has given rise to the tradition of *hanami*: blossom-viewing. It can be traced back to fertility rites and farming, like many other social traditions: when the cherry blooms it is a sign (from the gods, if you like) that it is time to plant your rice (Chapter 17). *Hanami* can be dated back to the Emperor Saga, who organised the first bloom-party in 812. This became an annual tradition, and spread out to the common people.

It is not just rapt gazing and poetic thoughts. It is also a good time, friends and families coming together, eating lovely food and knocking back the drink with a good will: a little like Christmas in Western cultures, which can also trace its origins to fertility rites. Picnicking groups gathered in parks inevitably merge into each other at the edges: and do so with good vibes because, like Christmas, this is supposed to be a happy time and one for sharing.

Under cherry trees
None are utter strangers

Another haiku from Issa. The difference between *hanami* and Christmas is obvious enough: cherry trees don't bloom on cue. That has led to the great Japanese tradition of tracking the blossom-front, called *sakura zensen*. The advance of the blooming from south to north is an abiding annual preoccupation. The blossom marches north from Okinawa, reaching Tokyo and Kyoto at the end of March, and eventually getting to Hokkaido in the deep north in May.

Most schools and many public buildings have cherry trees planted outside: for their brief annual beauty, for the reminder that beauty itself is brief, and because cherry trees are important to the whole idea of being Japanese. The Japan international rugby team is called the Brave Blossoms. This sounds effeminate to Western ears, but in Japan it means that they are aware of the fleeting nature of life and the need to seek glory, heedless of lesser concerns like personal mortality.

In the 1930s a secret society of young officers in the Imperial Japanese Army was formed; it was devoted to the creation of a totalitarian government. They called themselves the Cherry Blossom Society or *Sakurakai*. Pilots on suicide missions in the Second World War painted cherry blossoms on their doomed planes. Before the great Battle of Leyte Gulf in 1944, by some criteria the largest naval battle in history, Japanese sailors were exhorted to 'bloom as flowers of death'. The last message of Japanese forces on Peleliu in the Second World War was '*Sakura! Sakura!*'

Washington DC is bright with cherry blossoms in the spring: the trees were the 1912 gift from the mayor of Tokyo. In the twentieth century Japanese flowering cherry – mostly the variety known as *somei-yoshino* – spread around the world. In Nancy Mitford's *The Pursuit of Love*, the heroine Linda remarks: 'The great difference … between Surrey and proper, real country is that in Surrey, where you see blossom, you know there will be no fruit… The garden at Planes will be a riot of sterility, just you wait.'

But let's let Onitsura, yet another haiku poet, do the summing up, reflecting on the ordinary nature of the marvellous and the marvellous nature of the ordinary:

When cherry trees bloom
Birds have two legs
Horses four

FORTY-FOUR

FLAX

And the eyes of both of them were opened,
and they knew that they were naked.

Genesis 3:7

A great deal of philosophy, a fair amount of science and immense quantities of theology have been devoted to the task of creating an adamantine barrier between humans and every other species in the animal kingdom. This has turned out to be treacherous country, requiring constant changes in the way we define such concepts as humanity, language, consciousness, emotion and a thousand others. Charles Darwin famously pointed out: 'The difference between man and the higher animals, great as it is, certainly is one of degree and not of kind.'

However, I believe I can state without fear of contradiction that humans are the only species that makes and routinely wears clothes. (Though there are crabs that adorn themselves with seaweed as camouflage.) Of the 7.5 billion humans currently living on the planet, all but a few thousand wear clothes most of the time. No one is quite sure when this change took place: estimates range from 500,000 years ago to less than 100,000. No one is sure what brought about the change, either: we are all as free as philosophers to speculate without data. Protection from the weather? Magic? Modesty? Vanity? Prestige? The enduring human passion for inequality might incline us towards the last of these, but there is no evidence to get in the way.

The first garments were no doubt animal skins and vegetation; readers of a certain age will recall Raquel Welch's fur bikini in the 1966 film *One Million Years B.C.* There have been attempts to date the history of clothes by studying lice: body lice, unlike head lice, can't make a permanent home on humans because we are insufficiently hairy; it's been speculated that we evolved hairlessness to avoid body lice and the potentially fatal infections that can result from a bite. But lice are perfectly capable of residing in clothing, helping themselves to a blood meal when the clothes happen to be next to human skin. Such studies have produced a date of 170,000 years ago for clothed humans, but, like everything else on this subject, it remains conjectural. Increasing sophistication changed the clothes that people wore. Clothes don't preserve well, but other forms of technology do: sewing needles 40,000 years old have been found. That implies use of textiles rather than leather.

Plants you can wear: The Flax Carder, *1932, by Giovanni Giacometti (1868-1933); she is seated by a jar of flax flowers.*

That brings us to flax, the plant from which we get linen. Twisted and dyed flax fibres were found in the Dzudzuana Cave in the republic of Georgia, dated to 30,000 years ago. Flax was certainly cultivated in the Fertile Crescent 9,000 years ago. The seeds of the plant – linseed – and the oil are also useful commodities; the oil can be used to treat wood; until recently it was important to treat cricket bats with linseed oil. It's possible, perhaps even likely, that flax was first cultivated to make linen, rather than for its other virtues.

Creating cloth by hand is long, difficult and tedious. Early technologies include awls, loom weights and spindle wheels. The process of creating linen cloth involves retting the harvested stalks of flax; that is to say, removing the inner stalk. This is done in a standing pool of warm water and is followed by dressing, removing the fibres from the straw, and finally spinning, creating a continuous thread on a spindle, which is also called the distaff. The word has come to mean anything to do with women or with females: when breeding racehorses, you contrast the bloodlines on the stallion side to those on the distaff side. That's because spinning was traditionally women's work: monotonous, but you could do it when pregnant, and – until recent times – women generally were.

Flax is a good material to work with: perhaps it was the obvious choice to use as a basis for textiles. The strands are long – a flax plant stands around 4 feet (1.2 metres) – and strong, two or three times stronger than cotton. A flax field in bloom with its blue flowers is an attractive sight, but that's by the by.

As human society developed, the need for textiles, the need for clothing, increased. In Ancient Egypt the dead were wrapped in linen as part of the process of mummification: a great deal of human ingenuity and human thought has been based around the idea of life beyond death. Such ideas were naturally religious and religion is usually administered by special people. Linen cloth was the only possible apparel for priests in Ancient Egypt, because it was associated with purity. But linen was an important commodity for many other reasons.

Garments made from a single piece of cloth are still worn widely today, taking the form of the sari and the dhoti in India, the sarong in some other parts of Asia, and the kikoy in a great deal of Africa. Ties, folds and arrangements vary from place to place, but a deft piece of handwork makes a fine and surprisingly secure garment.

Linen grew increasingly important. The Phoenicians traded it; the Romans made sails from it. It went into decline when woollen goods became more prevalent, but the industry was revived by Charlemagne and was important through the Middle Ages. Linen was introduced to the Americas by the colonialists, and until cotton (see Chapter 74) took over in the nineteenth century, Europe and North America were largely dependent on flax for plant-based cloth. The crushed seeds that remain after the extraction of oil make an excellent protein-rich supplement to animal feed.

Cloth manufacture, almost inevitably, became the first process to become wholly mechanised. Industrialisation began with the weaving of textiles; it is also worth noting that the Jacquard loom, invented in 1804, could be programmed to make different kinds of cloth, so this was the first step towards the computer. Cloth is so important that it revolutionised human life at least twice.

The term linen is still used for items that are now seldom made from linen: as a generic term for underwear, which is kept in a linen drawer, and for bedding, which is 'bed linen' even when made from cotton or artificial fibres. Linen is still used for tablecloths and towels and for garments; when I was travelling to warm countries as a journalist I habitually wore a linen suit; the advantage is that it's *supposed* to look rumpled.

In the early days of civilisation linen gave those who wore it dignity and prestige. Perhaps it also gave us modesty, another concept shared by no other species. Either way, linen played a significant role in what defines us as human.

FORTY-FIVE

INDIGO

She comes in colours everywhere

'She's A Rainbow', Mick Jagger/Keith Richards

If you're going to clothe yourself in plants and you want to be as beautiful as the plants that surround you, you must borrow the colours of plants. You can colour the clothes you wear; you can even colour your skin. We have been dyeing our clothing for millennia, often with vegetable dyes. One of the most significant is indigo.

Indigo is the name of a dye and a colour as well as a plant. Two plants that contain indigo have played important roles in human history: the plant with the English name of indigo (because it originally came from India) and woad. Woad has a notorious meaning in the UK because it was used as body-paint in battle. Queen Boadicea – more correctly Boudicca – of the Iceni tribe of East Anglia painted her face with woad when she did battle against the Romans; in Scotland the Pict warriors painted their whole bodies with the stuff. The word Pict is from the Latin *picti*, which means painted. Julius Caesar wrote: 'All the Britons stain themselves with *vitrum*, which gives a blue colour and a wild appearance in battle.' There are continuing disagreements about what he meant: *vitrum* literally means glass. Woad is usually accepted as the right answer; you can then discuss whether the indigo it contains was used as a body-dye or as a tattoo. There are advantages to the use of woad in battle beyond giving yourself a wild appearance: it is also a powerful astringent. Vats of the stuff were prepared before battles to dress wounds.

Woad is found naturally east of the Caucasus, an erect perennial plant that grows to 4 feet (1.2 metres) in height. It came to Europe very early: it has been found in Neolithic and early Iron Age sites. It was used by the Ancient Egyptians to dye the linen used for wrapping mummies. Dyeing is not a straightforward business: no doubt it took years if not centuries to refine the process. A dye is different from a pigment: a dye binds to the material and gets right into it, instead of just colouring the surface.

If you want to make a dye from woad, you must harvest the leaves, wash them, cut them and then steep them for ten minutes in water at 80°C (175°F). You must then cool them very rapidly: an ice bath, usually unobtainable to the first users of the dye, is a great help in this. If you omit this stage, the dye won't go blue. After that, you strain your mixture and then add washing soda (soda ash) that has been mixed with boiling water. Whisk it for ten minutes and then store it in a jar for several hours. After that, strain it through cheesecloth and keep the sediment; you can dry it for later use, or use it at once. When you do so, add water and ammonia (human urine was used for this process) to the powder, heat to simmer and then immerse fabric in the dye. Bring it out and your fabric is a new colour: greeny-yellow. It turns blue as it meets the oxygen in the air. Do this all over again until you get a colour as deep as you want. This process was used in Europe for centuries: woad was a colour of choice as well as an ancient war paint.

The indigo plant – as opposed to the woad plant – had been used for dyeing garments in Asia for many years; it was cultivated in the Indus Valley 5,000 years ago. It came to Europe with the opening of maritime trade in the sixteenth century and met instant opposition. It was more efficient than woad and was in direct competition with the traditional dyeing industry. Governments responded with ferocious protectionist legislation: indigo was made illegal in Germany in 1577; use of indigo was made punishable by death in France in 1609; it was banned in England in 1660 because it was mendaciously said to be poisonous.

Labouring for colour: Indigo plantation, from Histoire Générale des Antilles Habitées par les François *by Jean Baptiste du Tertre (1610-87).*

But these strictures were mostly ignored, not least because it's impossible to tell which plant has been used to colour a garment. A little later Indian indigo gained a bad reputation, because it was often adulterated to make it go further. Plantations were set up in the French West Indies, which could supply the valuable substance relatively cheap because they used slave labour. When England and France were at war, indigo plantations were established in the United States, supplying England as well as the home market. An entrepreneur named Levi Strauss used the dye for his riveted work trousers in San Francisco; his company, set up in 1853, makes Levi jeans to this day.

Indigo dye creates a deep blue, and with repeated dyeing, it becomes almost black. It was used to dye English sailors' uniforms the colour known as navy blue; it was used to dye police uniforms. Why, then, do most of us think indigo is a sort of mauve? When Isaac Newton used a prism to break white light into its constituent colours, he badly wanted to see seven colours. After all, seven is a number heavy with significance: seven planets of the ancient world, seven days of the week, which represent the seven days of the creation, seven last sayings of Christ on the cross, seven deadly sins, and on and on. Naturally there were seven colours in the rainbow, there had to be: so Newton saw a colour in between blue and violet and called it indigo... which is by no means the only colour you can obtain with indigo dye.

Indigo production in India was about coercion. Indian farmers were forced into growing indigo under a debt system, and they got very little from the deal: the profits went to the British indigo planters, who didn't do any actual planting themselves, and the East India Company, which transported and sold the stuff. Two years after the Indian Revolt of 1857 (still in the UK more usually known as the Indian Mutiny), there was another rebellion, sometimes called the Indigo Revolt. The farmers refused to carry on planting indigo, saying they would rather beg. Soldiers were brought in to quell the violent uprising. The British rulers, nervous after the events of two years before, set up the Indigo Commission, under which farmers were required to see out their contracts with the indigo planters – but need not accept a repeat. It was a significant victory.

By the beginning of the twentieth century, however, indigo no longer mattered as much. Synthetic dyes could do the job much cheaper, if not better. Rebellion, slavery, the death penalty, exploitation and cruelty had all been considered acceptable in pursuit of our need to wear blue clothes. Now indigo is a curious bush, remarkable mainly for its history. But in recent years there has been a bit of a re-think: in South and Southeast Asia, indigo has made a comeback in textiles of high craftsmanship, the sort of thing sold to collectors and discerning tourists.

FORTY-SIX
GINSENG

I swear, calling upon Apollo the physician and Asclepius,
Hygeia and Panacea and all the gods and goddesses
as witnesses, that I will fulfil this oath...

The Hippocratic Oath

Perhaps the greatest human conceit is the idea that rational thought dictates our actions. We look at the damage we have done to our life-support system – the planet – and tell ourselves that at least we did what seemed at the time to be best for ourselves: creating wheat fields for our bread, cattle ranches for our meat and cities for our dwellings. We may be selfish, but we have been *logically* selfish: committed to rational actions which best advance our interests as individuals and as a species.

It's not a stance that can bear much looking at. The annual value of the global trade in ginseng has been calculated at more than US$2 billion. This is a commodity whose value to human health remains at best unproven. The value of the plant rises the more the bare, forked roots resemble a human figure: perhaps one equipped with a giant phallus.

Ginseng has been collected for thousands of years, in Asia and in North America. It grows wild in both places, a forest plant that does well in cool conditions. It is also widely cultivated, though this is difficult because the plant needs shade. It's a perennial herb with a stout taproot, inconspicuous flowers and red berries (drupes).

In Asia ginseng does best on the Korean peninsula and in northeast China. The word comes from the Chinese *ren sheng*, which means root of heaven. The discovery of its virtues is attributed to the legendary Emperor Shen-Nung, the Divine Husbandman and father of Chinese medicine. He tasted and tried more than 365 herbs, in some stories dying from the effects of the 365th. He chewed ginseng root and felt a pleasant sexual stirring and promptly recommended it for erectile dysfunction. This can be traced to the doctrine of signatures (see Chapter 5 on willow): if it looks like a penis, it's obviously good for the penis.

Opposite *Miracle plant: Mountain troll (samsin) with ginseng root; Korean ink drawing (undated).*

But the virtues claimed for ginseng go way beyond a sexual quick fix. The plant is said to boost energy, lower blood sugar and cholesterol, reduce stress, promote relaxation, operate as an anti-inflammatory and as a help with diabetes. It is said to work against physical and mental fatigue and to aid patients recovering from cancer. It is an immediately beguiling answer to the widespread twenty-first-century problems traditionally abbreviated by note-taking doctors as TATT: tired all the time.

Ginseng is obtained from species belonging to genus *Panax*. The name was given by Linnaeus, who knew of the claims made for the plant: it means cure-all in Greek, and is related etymologically to the term panacea; Panacea was a Greek goddess whose name means all-healing. The search for the panacea (sometimes tautologically referred to as universal panacea) was one of the goals of the alchemists, along with the philosopher's stone, which turned lead into gold, and the elixir of life, which conferred immortality on its users.

Alchemy was a mixture of experimental science and religious assumptions. The notion of a cure-all has a firm grasp on the human mind: it has driven the development of modern medicine and all of those who read these pages have cause to be grateful for that: most of us beyond a certain age owe our continuing existence to modern medicine.

I recall an argument between colleagues when I lived in Hong Kong. It involved a Chinese designer and an American editor: one maintained that traditional Chinese medicine was valid and important, the other that it was all rubbish and Western medicine was the only thing. The Chinese man held the latter view.

This is a polarising discussion: at its most extreme, total deniers line up against woo-woo merchants. It is as much a matter of temperament as of evidence: those who won't accept anything without a scientific seal of approval and those who point out that what's accepted and what's refuted by science changes every couple of years and that if we only accepted stuff we are capable of understanding fully we would seriously limit the possibilities of life.

This is all complicated by the placebo effect: placebo meaning 'I will please' in Latin. It's not a simple matter. The traditional sneer states that if you think it's going to do you good, it might actually do you some good, thereby proving (a) your ailments are imaginary and (b) you're stupid. But placebos can be effective even when the patient knows the medication is inert. The so-called open-label placebo – coming, for example, with the statement 'this is a placebo' – has been known to effect cures, genuinely reducing depression and perception of pain. Proponents say the process establishes a stronger contact between body and mind. In other words, ginseng can be very effective: its virtues have been summed up as a combination of coffee, Prozac and Viagra.

It's certainly highly valued. In the sixteenth century it was widely used in China for chronic illnesses and convalescence. It's grown in Canada and the United

States; Daniel Boone, the American pioneer and folk hero, made serious money from ginseng. Ginseng remains a popular universal treatment all over the world. It can be taken in the form of tea made from the leaves; at the 1988 Olympic Games in Seoul, South Korea, the British sprinter Linford Christie failed a drugs test, but was let off when he explained that he had been drinking ginseng tea. (Christie was banned for two years after failing another drugs test some years later.)

Ginseng can be served in food, usually as an ingredient in side dishes in Chinese and Korean cuisine; it's also made into an alcoholic drink. Some of the most anthropomorphic (and perhaps ithyphallic) roots are framed and given as presents. A price of US$500,000 has been claimed for one such. Wild ginseng fetches much higher prices than the farmed product, even though, in terms of phytochemicals (in this case ginsenosides), there is no measurable difference. A study in South Korea in 2002 found that 60 per cent of men treated with ginseng found improvement in their sexual performance.

Is this rational? Is rationality even a relevant concept? I am reminded of the author's foreword to *The Collected Poems of Dylan Thomas*: 'I read once of a shepherd who, when asked why he made, from within fairy rings, ritual observances to the moon to protect his flocks, he replied: "I'd be a damn' fool if I didn't."'

FORTY-SEVEN

RUBBER

Like a rubber ball I'll come bouncing back to you

'Rubber Ball' by Jimmy Cliff, sung by Bobby Vee

It's probably the oldest of all ball games, and certainly the oldest played with a rubber ball. It's usually referred to as the Mesoamerican ball game, and it was played at least as early as 1600 BC. It was continued across the centuries by the Olmec, Maya and Aztec people. No one is really sure of the rules or even how it was played, though human sacrifice is said to have been involved, whether for the winning or the losing team is unknown. At times in history it was played with a ball that weighed 9 pounds (4kg).

The rainforest tree *Hevea brasiliensis* exudes a milky liquid, latex, to heal its own wounds. It's far from unique; other plants produce a similar substance. This can be used to make rubber. The Aztecs not only made balls; they also used it to make containers for liquids, and they impregnated cloth with rubber to make it waterproof. Latex can be obtained from the trees by wounding them deliberately and using a cup to collect the liquid, which solidifies into lumps. These lumps have remarkable qualities: stretchy but strong, as well as waterproof.

European explorers learned about rubber from the indigenous people and by the eighteenth century they were exploring its extraordinary properties. But it was a bit like silly putty: for all its marvellousness, it was hard to put to practical use. Waterproof shoes were tried, but the stuff is extremely sensitive to changes in temperature; shoes that crack when too cold and melt when too hot are of limited use. In 1770 Joseph Priestley found that it worked as a pencil eraser: handy, but not world-changing.

An American inventor named Charles Goodyear did the changing. He was a self-taught chemist, engineer and entrepreneur and was never the sort to die wondering. He believed that rubber had enormous commercial potential if only he could come up with the right treatment. He had already experimented with a colleague on an unsatisfactory mixture of rubber and sulphur. But in 1839 came one of those glorious accidents of the penicillin kind: Goodyear dropped a quantity of rubber treated with sulphur onto a hot stove and that smoking lump was exactly what he had been looking for: a piece of material that was pliable, waterproof and

Opposite *World-changer: rubber shown on a French educational card, late nineteenth or early twentieth century.*

EL CAUTCHUCO — LE CAOUTCHOUC — THE INDIA-RUBBER

mouldable, yet still strong. Crucially, it was not affected by extremes of cold and heat. This was the process called vulcanisation, named for the Roman god of fire.

Goodyear took out patents, but they weren't valid in Europe and anyway his work was pirated in the United States. He died in immense debt in 1860, a couple of decades before the beginning of the great rubber boom that he made possible. This came first with bicycles and then with motor vehicles. From the late nineteenth century the world was mad for rubber. The stuff had other important industrial uses – hoses, conveyer belts, gaskets, matting and flooring. The world's only source of rubber was the Amazon, and this explosion of demand was exploited by the rubber barons of Manaus in Brazil. Legends of their wealth and flamboyance are still told: lighting cigars with $100 bills, watering their horses with champagne and establishing the most lavish brothels on earth. They gathered the rubber from widely separated wild rainforest trees with captive workers kept in order by private armies.

But soon others were competing for the same market. Leopold II of Belgium acquired the Congo Free State as a personal possession after he convinced the Berlin Conference on Africa of 1884-85 that his intentions were wholly philanthropic. He then established the practice of collecting rubber from the jungle vines (called Congo rubber) with forced labour. He kept his workers in line with systematic killing: the hands of victims were chopped off as visible proof and to keep the rest from slacking.

In 1876 Henry Wickham of England smuggled 70,000 seeds of the rubber trees from Manaus and took them to the Royal Botanical Gardens at Kew, near London. Of these, 2,400 germinated and seedlings were sent out to British outposts in what is now India, Sri Lanka, Indonesia, Singapore and Malaysia. The industry was promoted by Sir Henry Ridley, first director of the Singapore Botanic Gardens, who was known as 'Mad Ridley' for his enthusiasm for rubber. He developed techniques of tapping the trees without causing lasting harm, and persuaded planters, comfortable with their crops of tea and coffee, to switch to rubber.

The motor industry expanded with immense rapidity, and it was able to do so because of the world's rapidly increasing rubber production. This accelerated after 1888, when John Dunlop, a Scot living in Ireland who owned a large veterinary practice, developed the first practical pneumatic tyre; he designed it for his son's tricycle. Rubber was now a crop rather than a wild resource; it was manageable and renewable. The headquarters of this industry shifted from South America to Asia, and that continues to this day. This had important tactical implications: rubber was now an essential commodity, so trade links with rubber-producing countries had to be maintained. Nazi Germany, nervous about losing such contacts, tried to produce useable rubber from dandelions; at around the same time, nervous of Japanese influence in Asia, the United States tried the same thing, both without success despite the promising qualities of the white liquid inside the plants.

A rubber tree takes seven years to reach an age at which it can be exploited, after which it will be productive for up to twenty-five years, though this depends on the skill of the tappers. Rubber trees do well in warm, wet climates with plenty of sun. Further uses of rubber have been found: medical gloves, elastic, as an ingredient in cement, and for condoms, used for birth control and, since the AIDS pandemic of the 1980s, increasingly for sexual hygiene.

Naturally there was a race to create synthetic rubber. It was first produced in 1910 and developed in subsequent decades. It is made from by-products of petroleum, so it is not a renewable commodity. The costs of synthetic rubber doubled overnight in 1973 after the Organization of the Petroleum Exporting Countries (OPEC) imposed an embargo on the world's supplies. A further boost to the natural-rubber industry was the switch to radial tyres, which give longer life, better steering and more economic use of fuel. With radial tyres the steel cords that give strength to the rubber are at 90 degrees to the direction of travel; this gives much better shock absorption but leaves them vulnerable in the sidewalls – when you bump a kerbstone, for example. Natural rubber is needed in the manufacture of such tyres; radials made from pure synthetics are not strong enough.

The formerly ubiquitous houseplant called the rubber plant is not closely related to the rubber tree. The pot-plant is *Ficus elastica*, which can grow to 130 feet (40 metres) in its native Asia. It has been used as a source for rubber, but these days its commercial use is restricted to decoration. In the meantime, the world continues to run on rubber, and a good deal of it still comes from trees. One billion tyres are sold every year to keep the world rolling. Rubber trees can suffer from a fungal leaf blight: it represents a considerable threat to the way the world has chosen to live. The infection causes defoliation and consequent inability to photosynthesise; it affected about one tenth of Indonesia's rubber trees in 2019. It exists as one more potential crisis.

FORTY-EIGHT

GARLIC

Without garlic I simply would not wish to live.

Louis Diat, French-American chef

For the Anglophone cultures of the world, garlic represents a classic contradiction: a deep satisfaction with everything you grew up with alongside a wild longing to escape it all. The same phenomenon exists in every culture: perhaps in every individual in every culture.

For the British, garlic was everything obnoxious about the far side of the Channel. In Nancy Mitford's *The Pursuit of Love*, Uncle Matthew, an English aristocrat, maintains: 'Abroad is absolutely bloody and all foreigners are fiends.' His daughter falls in love with a French duke and finds utter bliss.

Percy Shelley wrote home in 1818: 'There are two Italies… the one is the most sublime and lovely contemplation that can be conceived by the imagination of man; the other is the most degraded, disgusting and odious. What do you think? Young women of rank actually eat – you will never guess what – garlick!'

Garlic was in fact used in British food across the centuries, even if it was not entirely respectable: Chaucer wrote of the hideous Summoner in *The Canterbury Tales*: 'Wel loved he garleek, onions, and eek lekes'. Despite the contradictions of the British attitude, in most of the rest of the world garlic has been grown and even treasured across the millennia, for the flavour it brings to food and for its medical benefits.

Garlic is a member of the *Allium* genus that includes onions, chives and leeks; the garlic species we mostly eat is *Allium sativum*. Garlic releases phytochemicals when it is damaged, as a defence mechanism against insects, worms and birds (we have met this phenomenon before in these pages). This is what gives the garlic its flavour. Chopping, chewing and crushing releases the well-known garlic odour. The flavour becomes a good deal mellower with cooking, where it gives background and depth rather than an absolute belt, apart from in dishes where it is meant to – for example *spaghetti aglio, olio e peperoncini*, spaghetti with garlic, olive oil and chillies.

Garlic is an ancient part of the cuisine of most of Asia and the Mediterranean, giving liveliness to the flavours all around it and also, of course, allegedly doing the

Opposite Good for you: garlic-growing, illustration from Tacuinum Sanitatis in Medicina, *the Latin version of an Arabic work on health and science (late fourteenth century).*

leum. ꝯplo. ca. m. iij. ſic m. iij. Et ect arbozes ex eo. quo é modice acutatis. uiuani. ꝯſ
enca. ſit. ꞇ mozſus ſcozpionu. ꞇ inpar ꞇ inſicit umes. ll ocumitr̃ nocet oc̃ul ꞇ ccirbzo.
emio nocti cu aceto ꞇ olo. Quo gn̄tur humoz ꝰ gꝛoſſis ꞇ acutis ꝯuenit ſit. decꝛepi
ꞇ ſcnib; hyeme ꞇ montanis ꞇ ſeptentrioalib; :

Good eating: shopping for garlic in Winter, 1595, by Lucas van Valckenborch (1535-97).

consumer all kinds of good. The two things – good taste and good health – have been part of the appeal of garlic throughout history, a win-double if ever there was one.

Garlic has probably been cultivated for at least 6,000 years; it is referred to in cuneiform tablets of Mesopotamia at least 4,000 years old. It was fed to athletes at the ancient Olympic Games before they competed, an early example of a performance-enhancing drug. It was given to slaves in Ancient Egypt, including or perhaps especially those that built the pyramids. Garlic was found in the tomb of King Tutankhamun; it is not clear whether it was placed there for ritual reasons or because one of the workers mislaid his lunch – but the tombs also contained vessels in the form of garlic bulbs, so perhaps the power and strength given by garlic was considered useful in the afterlife.

Pliny the Elder was full of praise: 'Garlic is very potent and beneficial against all ailments caused by changes of water and of location; it drives away snakes and scorpions by its smell and, according to some authorities, every kind of wild beast.' Garlic will repel mosquitoes and slugs; in Romania it was supposed to keep vampires at bay. The Greek physician Hippocrates prescribed it for respiratory problems, parasites, digestive problems and fatigue. The great Hildegard von Bingen, the twelfth-century saint, abbess and philosopher, best remembered as a composer, talked up garlic in her medical writings. In the sixteenth century the German botanist Adam Lonicer said that garlic was 'the rustic's theriac', or cure-all. It is said to lower blood pressure, cure colds, reduce levels of cholesterol, reduce risk of heart diseases, stave off dementia and improve athletic performance. In India it was considered to be an aphrodisiac.

Perhaps for that reason, Buddhists avoided it. It's acceptable in Islamic traditions, so long as you don't consume it before going to the mosque, which is presumably as much an act of politeness as religious observance. It was probably the Muslims who spread garlic into Europe, where it was considered to have benefits against smallpox. Louis Pasteur discovered its antibiotic properties in 1858; it was used, apparently with some success, during the cholera outbreak of 1913 and the so-called Spanish Flu pandemic of 1919. When supplies of penicillin (discovered in 1928, see Chapter 21) were running short during the Second World War, the Russians used garlic to treat wounds; it was known as 'Russian penicillin'.

Modern Western medicine is traditionally resistant to folk remedies of all kinds, though that instinctive rejection is now changing in many places. Scientific tests and surveys on the use of garlic have indicated – it would be rash to use the word 'proved', and scientists hate rashness – that garlic can be helpful with cancers of the lungs, brain and prostate, also with osteoarthritis. It has genuine antibiotic properties. It can help with problems of the heart and the blood pressure. It might protect the liver from damage from alcohol consumption. It might lower the risk of premature birth, and it may help prevent colds.

The health benefits of garlic were always an attraction, but the fact that garlic adds and enhances flavour in cooking is just as important. The downside is that it taints the breath: an old Jewish joke says, 'A nickel will get you on the subway but garlic will get you a seat.' Parsley is supposed to lessen or even negate that effect, which is one reason (other than excellence of flavour) for pairing the two, in garlic bread, escargot and other dishes.

Garlic was rejected strongly by the British until the 1950s, when the country began to emerge from post-war austerity and people sought a less circumscribed life. Garlic became acceptable largely because of Elizabeth David, an upper-class rebel who studied art in Paris, worked as an actress, ran away with a married man, sailed to Italy in a small boat which was then confiscated, reached Greece, narrowly escaped from the Nazis and got to Egypt. In 1950, back in England, she published *A Book of Mediterranean Food*, a well-written, effortlessly cosmopolitan work in which she introduced the British to aubergines, basil, olive oil and, of course, garlic. In this and subsequent works she was firm in permitting no reckless substituting of ingredients in classic recipes, and yet was never over-elaborate. You could cook like this; anyone could. Many of these items were hard to obtain in Britain: it was cooking as adventure, cooking as aspiration, cooking as a form of courage, cooking as a portal to a new world.

My mother was a wholesale convert. Her own mother told her this fancy foreign cooking was all very well, so long as she didn't use garlic. Unknowingly, my grandmother tasted and praised many garlic-infused dishes when seated at the big table in Streatham.

FORTY-NINE
FOXGLOVE

…The heart-ache and the thousand natural shocks
That flesh is heir to…

William Shakespeare, *Hamlet*

Herbal medicines occur over and over again in these pages: many plants have been considered to have beneficial effects on the human body. But it's not wise to get superior about it. Just in case recent chapters on ginseng and garlic imply a negative attitude to such remedies, we'll turn to the foxglove. We have

For medical use: Portrait of Dr Paul Gachet, 1890, by Vincent van Gogh (1853-90); he is holding a foxglove.

already discussed the fungus *Penicillium* that gave us penicillin (Chapter 21) and the willow that gave us aspirin (Chapter 5): never forget that roughly half our medicines, including many of the most important, are taken from sources in the natural world, most of them plants and fungi.

There's another thing to consider here: the idea that most herbal remedies are pretty harmless, so you can hand them out in vast quantities on a hit-or-miss basis. Not a great idea with foxgloves: a lovely plant that will kill you if you ingest enough of it.

There are about twenty species in the genus *Digitalis*, the foxgloves. The Latin term comes from the word for finger, because you can wear a single foxglove flower on your fingertip. They are familiar to many of us: biennial plants that grow a good four feet (1.2 metres) in height and bear spectacular clusters of bell-shaped flowers in purple, pink, white and yellow. Cultivated varieties are grown in gardens. The name is a tease: there's an enchanting notion that people believed foxes slipped the flowers over their paws to aid silent hunting. A more likely notion is that they were actually called folk's gloves. The folk in question are little folk, fairy folk, but they are not to be named as such, because saying their name out loud will attract their attention and bring bad luck: you must never name *them*. The marks inside the bells are said to be the handprints of fairies. Foxgloves have also been called dead man's bells, for the poison they contain, and witches' gloves for their association with witchcraft.

The plant was first described in 1542 by the German botanist and physicist Leonhard Fuchs (from whose name we get fuchsia). But a greater scientific interest in the plant had to wait for another couple of centuries, when William Withering used substances gained from foxglove plants to treat a patient with the ailment that was then termed dropsy. This is the swelling that is caused by congestive heart failure, now called oedema. It worked. He recognised that foxgloves provided the active ingredient in the cure and wrote it up, for he was a Fellow of the Royal Society. In 1785, he produced *An Account of the Foxglove and Some of its Medical Uses*. The same year, Withering was called in by Erasmus Darwin, physician and grandfather to Charles, on a similar case, and the same treatment was successful. Darwin then produced a treatise of his own: *An Account of the Successful Use of Foxglove in Some Dropsies and in Pulmonary Consumption*. This led to a bitter row about priority, and the two men became estranged. Nevertheless, the beneficial effects of foxglove were now in the public domain.

Digitalis, the substance obtained from foxgloves, contains several compounds. These have since been separated and one of them, digoxin, is used to treat heart complaints and to control irregular heartbeats. It works as a muscle stimulant, and since the heart is a muscle, it affects the heart. A moderate dose will slow the heart, replace an irregular beat with a regular one, and make each heartbeat more forceful and effective. It has been used to treat high blood pressure; it was also used to treat epilepsy, but no longer.

The entire plant is toxic, and that makes it somewhat crucial to get the dose right: there is a very small gap between a therapeutic and a harmful dose of digitalis. An overdose will cause nausea, vomiting, diarrhoea, and an abnormal heart rate that can lead to collapse and death. It can also cause hallucination and delirium and can affect colour vision. It can give perceived objects a yellow tinge, and cause a halo to be seen around sources of light. It has been speculated that Vincent van Gogh was given digitalis therapy to control the epileptic seizures he suffered from: and that this led to what's been called his Yellow Period. Perhaps that also explains the extravagance of his great *Starry Night* painting of 1889. There's a clue: in his portrait of Dr Gachet, the subject is holding a foxglove… but van Gogh didn't meet Gachet until May 1890. It remains speculation; and anyway, what matters is the paintings.

There's an unanswered question hanging over this account of the rise of digitalis: how did Withering know that foxgloves would work? How did he even know that it was worth trying? How did he come to use a potentially lethal substance as part of a cure – and a successful one at that? One answer: he was given the recipe by a wise woman by the name of Mother Hutton. This story has great appeal: it was meant to. It was a later invention, used to advertise the drug's effectiveness; Withering never mentioned such a person. She was invented in 1928 by the drug company Parke-Davis.

But the fact remains that Withering must have got the idea from somewhere – and that moves us into vague and dangerous areas of history. Foxgloves were certainly used to treat dropsy before Withering got hold of the idea, but it is part of the shadowy history of folk remedies, herbal medicine and what we now call superstition. Until recent times there was no clear distinction between science and magic, while many unprovable beliefs were considered irrefragable facts.

The history of folk medicine is tied up with the tradition of the wise woman: the person who understood the secrets of plants. This was of course connected to belief in witchcraft. The persecution of witches is ancient; when things go wrong it must be somebody's fault. It follows that experts in herbal medicines were persecuted as witches, and many were executed: it's been estimated that in Europe, in the three centuries to 1750 – that's the eighteenth century, the century of the Enlightenment – between 40,000 and 100,000 people were executed for witchcraft.

It is, then, hardly surprising that precise records of the use of natural ingredients in remedies are confused and vague. Meanwhile, herbal medicines have had a revival in the past half-century or so. And at the same time, countless people have benefited from foxgloves, as used in conventional medicine.

FIFTY

GRAPE

With thy grapes our hairs be crowned:
Cup us: till the world go round!

William Shakespeare, *Antony and Cleopatra*

When I complete this chapter I will have reached the halfway point of this book. Shall I reward myself? Shall I mark this semi-achievement with a treat? If so I must surely turn to the grape. Grapes grow in bunches of between fifteen and three hundred berries. The skin of grapes is very attractive to air-borne fungi known as yeasts (see Chapter 11); so much so that you can use grapes (I have done so) to start the process required to make a starter for sourdough bread (see also Chapter 2 on wheat). But that's not the main use we humans make of the grape's readiness to ferment.

Jamshid, the legendary king of Persia, was a great lover of grapes, which he consumed from a special jar. But one time the grapes were kept too long and were horrid. They were marked as poison and set aside for future use – a rather sinister thought. A woman who had been banished from his harem found it and drank it in her despair. But instead of dying, she felt better than ever before. She told the king. He was delighted by her discovery and promoted her back to the position of royal favourite. Wine had been invented.

Grapes are high in sugar as well as being magnets for yeast, some varieties as high as 24 per cent. The yeast converts the sugars to ethanol, carbon dioxide and heat. There is evidence for wine manufacture dating back 9,000 years in China, 8,000 in Georgia, 7,000 in Iran, 6,000 in Sicily. Wine spread from the Caucasus – what is now Armenia, Georgia and Azerbaijan – into Europe and North Africa with the Phoenicians. The West Asian heritage was celebrated in *The Rubaiyat of Omar Khayyam*:

Ah, with the Grape my fading life provide,
And wash the Body whence the Life has died,
And lay me, shrouded in the living Leaf,
By some not unfrequented Garden-side.

Wine was important in Ancient Egypt: thirty-six wine amphorae were found in the tomb of Tutankhamun. The Ancient Greeks were immensely keen on wine: in

Joy at a price: The Young Sick Bacchus, *1593-94, by Michelangelo Merisi da Caravaggio (1571-1610).*

The Odyssey the hero often seems to navigate from one booze-up to another across the wine-dark sea: one of Homer's favourite phrases. 'It is the wine that leads men on, the wild wine that sets the wisest man to sing at the top of his lungs, laugh like a fool – it drives the man to dancing – it even tempts him to blurt out stories better never told.'

When did wine become important in religion? Perhaps ever since its invention, as an aid to ecstatic experiences. The Greeks celebrated Dionysus as the god of wine, who became the Roman god Bacchus. The Romans planted vineyards near garrison towns: long before Napoleon spelt out the principle, they knew about the importance of morale in an army. In Judaism the practice of Kiddush, the blessing of wine (in abstemious households the unfermented juice of grapes) for Shabbat is ancient. It is the origin of what is known as the blood libel: the Christian idea that the Jews habitually murder Christian children to use their blood in rituals.

Wine has been central to the daily (Good Friday excepted) Catholic ritual of the Eucharist for 2,000 years, tracing its origins to the Last Supper, when Christ offered wine to his disciples saying: 'This is my blood.' It follows that wherever Catholic conquerors and missionaries travelled, they established vineyards: if you are not receiving the sacraments you are not living a Catholic life. Once wine was established, its secular uses could also be savoured.

It's easy to make wine; it's hard to make good wine. Nature's tendency to turn everything into alcohol is all very well, but the production of drinkable wines in quantity requires a little sophistication. The fundamentals, however, are simple enough. To make red wine you harvest your dark grapes and crush them. After that you let them ferment. After a well-judged amount of time – say five to seven days – you press the grapes, which removes the pulp, the pips and the skins. The remaining liquid is matured in oak barrels or stainless-steel tanks. Then it's usually blended with other wines until you get the taste you want, and it's bottled, sold and drunk. The process of bottling generally involves fining, to improve the clarity. The material used for this is often from animals, including gelatine, egg white and casein (from milk), which makes many wines unsuitable for strict vegans. These days a lot of wine is sold in bottles with screw tops. More prestigious wines are still sold in corked bottles: contact with the cork and with the air improves their quality. That's why such wines are laid down to rest for a few years so that they can be consumed at their peak.

White wines can be made from black grapes: it's all a matter of when you do the pressing, the process that removes the skins. The skins that add flavour and tannins to reds are not wanted after the initial stages in whites. To make white wine you press before you ferment. To make rosé wines you ferment at a cool temperature, as you do for most white wines, and do so only for a brief period before pressing.

France remains top of the per capita world wine consumption, with Italy just shading it in acreage of vineyards. This gives rise to what's called the French paradox: the French diet is high in animal fats, but French people have a relatively low level of heart disease. This is usually explained by the health-giving properties of wine, though it's also been attributed to the way the French collate their health statistics.

There is more than one species of grape: some members of the *Vitis* genus are native to North America. This simple fact led to the devastation of the European wine industry in the nineteenth century. Phylloxera, a species of aphid that feeds on the roots of vines, was accidentally introduced to Europe via the importation of infected American vines. These spread into European vines, which are a different species with no natural defences against the aphids. (American species develop a sticky sap that clogs the aphids' mouthparts.) Nobody starved to death as a direct result of infected vines, but many businesses were destroyed. European vineyards were rescued by grafting: creating hybrids with American species.

Grapes are also grown for eating. These days many are seedless, which is as unnatural a practice as can be imagined; the plants are propagated by cloning. About 71 per cent of the world's grape production is for wine, 27 per cent for eating fresh, and what remains is dried.

Wine is prone to all kinds of snobberies and nonsense, inevitably more in England than anywhere else, where the upper classes traditionally drink wine and the lower classes beer, though the process of democratisation can now be observed in every supermarket. This is a comparatively recent development: in childhood holidays in Cornwall the only shop in the nearest town that sold wine was the chemist. Even thirty years ago it was mildly unusual to find wine in a pub: now you are usually asked to specify which grape variety you prefer.

So let us savour the conversation between Charles and Sebastian in Evelyn Waugh's *Brideshead Revisited* as they ransack the family cellars:

'…It is a shy little wine, like a gazelle.'
'Like a leprechaun.'
'Dappled in a tapestry meadow.'
'Like a flute by still water.'
'…And this is a wise old wine.'
'A prophet in a cave.'

But as I approach the halfway point of this book my thoughts turn to champagne. It is pleasant to reflect that bubbles were once the shame of the Champagne region and they did everything they could to get rid of them. This is the most northerly wine-growing region of France, and in cold weather the fermentation stops as the yeast goes to sleep. When it warms up again the yeast wakes up and causes fermentation in the bottle, creating a lot of carbon dioxide – so either the bottle bursts as a result of the pressure or the wine goes fizzy. Dom Perignon himself did his best to get rid of the bubbles. But some rather liked them. The British took a shine to the stuff; after all, this is the closest French wine to the Channel ports. After the death of Louis XIV in 1715, the French court took to champagne. For a while the fizziness remained something of a chance matter, but in the nineteenth century they established better control over the process and had stronger bottles to put the wine in. The *methode champenoise* was developed by Veuve Clicquot, and the major champagne houses were established in the nineteenth century: Bollinger 1829, Krug 1843, Pommery 1858.

Champagne is the best drink for celebration purposes for reasons of biochemistry as well as for its reckless pricing. Fizz – bubbles of carbon dioxide – pushes the alcohol into the bloodstream at the fastest possible rate: that's why so many alcoholic drinks (lager, whisky and soda, gin and tonic) are fizzy. Champagne has the optimal percentage of alcohol to facilitate that process. And you, dear reader, if you have been with me from the beginning of this book – perhaps you too deserve a glass of champagne.

FIFTY-ONE

EDIBLE MUSHROOM

…and you whose pastime
Is to make midnight mushrumps…

William Shakespeare, *The Tempest*

In a way all mushrooms are magic mushrooms. They don't behave in a way that makes easy intuitive sense: they grow in the dark, they pop up overnight, they don't have flowers or seeds, there's not a speck of green on them but they grow and they vanish, and often turn up again the following year as mysteriously as ever. What's more, some of them can send you off your head and a good few can kill you.

You can find edible mushrooms on the same supermarket shelf as vegetables, but they're not plants. We're prepared to eat them with good appetite, but comparatively few people in Anglophone cultures are willing to gather and eat wild mushrooms. Mushrooms are good – but you can't trust 'em. The relationship between humans and mushrooms has always been rather fraught.

Eat this painting: still life with a plate of azaroles (berries of a hawthorn species), fruit, mushrooms and cheese (undated) by Luis Egidio Meléndez (1716-80).

It's been estimated that of the readily available species, 50 per cent are inedible, being woody or otherwise unsuitable; 25 per cent are edible but flavourless; 20 per cent will make you ill in one form or another; 4 per cent are edible and tasty and 1 per cent can kill you. Now there are plenty of inedible plants, plenty of plants that make you ill and a good few that will kill you: but the same blanket suspicion of fungi doesn't operate for plants. I suspect that's because fungi fuddle the mind long before we eat them. We expect them to behave like plants, but they don't, because they're not. As we have seen, they belong to a separate kingdom: equal in status but for ever different to us animals and those plants.

Fungi don't get their energy directly from the sun. They get it second-hand, as we animals do: by consumption. That's why fungi grow on trees or, more disturbingly, on animals. We like the world to be binary: we naturally and intuitively divide living things into plants and animals: this third group disturbs our sense of order. Fungi are neither the one thing nor the other: they are utterly and rather dismayingly themselves.

But they are also a readily available source of nutrition and humans have been gathering and eating mushrooms for millennia. No doubt our hunter-gathering ancestors had fungi in their diet. There are signs of mushroom consumption at a site in Chile 13,000 years old. The Chinese tradition of eating mushrooms goes back thousands of years, often linked with a belief in their medicinal and therapeutic qualities. In Ancient Greece and Rome, mushroom consumption was associated with the elite: if you could employ a food-taster – not the most attractive vocation – the dangers inherent in mushroom consumption could be minimised.

The wariness about fungi is based on common sense, but it varies wildly from place to place. In many cultures autumnal fungus-collecting trips are a family ritual. I used to visit a café in Trent Park, a woodland at the end of the London Tube line. It was run by an Italian family: in autumn they served *pasta ai funghi*, gathering the fungi that the poor Brits were too terrified to touch. It's a process that produces a great feast and also, since many species can be dried, an opportunity to store tasty food for the winter. The traditional way of doing this is to slice the mushrooms thinly and hang them together on a thread. More than 90 per cent of a mushroom is water; drying intensifies the flavours. You can also do this in an oven: twenty minutes each side at 65°C (150°F) should do it.

The gathering of wild mushrooms is ancient, but their cultivation is not. It can be difficult and time-consuming, and for as long as there were wild woods and an underclass to do the gathering, there was no need. But in the seventeenth century in Paris the demand for mushrooms created a need for nearby urban sources. Louis XIV had a taste for them, so there was obvious profit in catering for it. Quarry workers noted that edible fungi – the kind we know as button, Portobello or Parisian mushrooms – grew spontaneously on the dung of the horses that transported the quarried stone. It was then that an ingenious person named Monsieur Chambry brought the dung heaps inside the tunnel of the quarries where the fungi could grow throughout the year. By the nineteenth century, when cement had made the quarries redundant, they became full-time mushroom farms and flourished until the Metro was built in the early part of the twentieth century.

Mushroom cultivation is now an established business. Its main requirement is a substrate on which the mycelia – the underground threads that constitute the main part of fungi – can feed and so produce the edible fruiting bodies. Suitable materials include woodchips, sawdust, mulched straw, horse and poultry manure, corn cobs, recycled paper, coffee pulp and grounds, nut and seed hulls, soy meal, brewers' grains and urea. These days production mostly takes place in purpose-built windowless buildings designed to keep out contaminants and allow easy harvesting. It's not a straightforward business: an acquaintance of mine produced mushrooms in quantity, but often they weren't of the uniform size and shape demanded by supermarkets. In those days I was always being given huge baskets of delicious but unsaleable mushrooms to turn into stroganoffs.

The English word mushroom is mostly used for a single species, *Agaricus bisporus*, native to Europe and North America. It is readily cultivated; varieties include cremini and Portobello, already mentioned. Other cultivated species include chanterelle, morel, oyster, porcini or ceps, shiitake and the wood ear, formerly known as Jew's ear. Foraged and cultivated edible fungi can be found in season in many European markets: up to 300 varieties have been claimed for the Viktualienmarkt in Munich.

In recent years the food giant Rank Hovis McDougall came up with a method of producing food from microfungi, retailed as Quorn, and described as mycoprotein. It came on the market after a twelve-year testing process for pathogens. It is grown in vats of glucose syrup, and the resulting food is high in proteins and low in fat. It is marketed as a meat substitute for vegetarians and vegans and is good at absorbing flavours: it works particularly well in Indian recipes heavy with cardamom.

Those from Anglophone cultures tend to regard eating any but the most obvious mushrooms as a daring event, one that shows our dashing cosmopolitan nature. Shirley Conran, in her guide to being a late-twentieth-century female *Superwoman*, famously declared that life was too short to stuff a mushroom (presumably she never tried; it's not a terribly fiddly or time-consuming job). Large open-cap mushrooms sometimes turn up as burger substitutes in pubs, as a vegetarian option that requires little preparation. Funghi porcini are best for pasta and risotto.

FIFTY-TWO

JAPANESE KNOTWEED

*'Would you like foreigners to come into your house, settle down
and help themselves to your fridge?'*

Jean-Louis Debré, French interior minister, 1997

Accounts of invasive plants always sound like science fiction: as if the plants were the creation of an evil genius or the emanation of a hideous force from a distant galaxy. The success of an invasive species seems both inevitable and a moral judgement on humankind. The fact is that despite the success of kudzu (Chapter 35), it's hard for any plant to establish itself in an alien environment. For a start there are, more or less by definition, very few of them in the new place, so the plant has nothing in reserve and few opportunities for sexual reproduction. There's a rough rule of 10 per cent: only about 10 per cent of plants that find themselves in an alien environment survive in the wild without human help. Of these, only about 10 per cent establish themselves as viable species. The process can be seen as the most searching possible examination of a plant's reproductive strategy.

An alien does start with certain advantages. There are no natural predators: no animals, fungi or plant parasites that have evolved to exploit the plant. There are no indigenous diseases to keep it in check. Nothing in the new environment has an ecological niche based on the damage it can do to any alien plant: it would be a contradiction. But that works the other way: there is nothing there to help it either: no natural pollinators, no natural dispersers of seeds. The conditions of the new country are unlikely to fit the plant's natural requirements in terms of temperature, moisture, soil type and seasonal rhythm. A successful invasive plant needs toughness and the ability to operate as something of a generalist. A specialist that has perfectly evolved for one type of environment is unlikely to be able to transport those skills to another. The all-rounder has a better chance.

Japanese knotweed fits that description pretty well, and it has become a serious problem in Europe and North America. It joins a long list. The Royal Botanic Gardens at Kew calculates that 6,075 global species of plants are invasive. Some of them are of minor significance, but others have had a dramatic effect on their new environment: as well as kudzu, there are problems with giant hogweed, Himalayan balsam, rhododendron, New Zealand pygmy weed, purple loosestrife, Norway maple, English ivy (in America), caulerpa seaweed, water hyacinth, tamarisk

Not such a challenge: cartoon by Glenn Marshall.

and many others. In Hawaii, the invading miconias are attacked with a paintball gun from a helicopter, the pellets full of herbicide. Koster's curse is an invasive bush now found in many tropical regions outside its Central American home range.

Japanese knotweed grows, as the name suggests, in Japan, as well as the Korean peninsula and northeastern China. Here it must cope with thirty species of insects and six of fungus that target it. The plant was much admired by Philipp Franz von Siebold, the German physician and botanist who introduced Western medicine to Japan. He thought that Japanese knotweed was a great boon to humanity and brought it back to Europe in 1830 as part of a huge collection of plants and animals. He settled in Leiden, where a botanic garden was established. He also distributed knotweed around Europe. He sold it to any takers as a fast-growing and ornamental fodder plant. Samples were sent to Kew and to Edinburgh; by 1854 Japanese knotweed was on sale at a nursery in Kingston, near London. The plant was naturalised in the UK by 1886.

Plenty more where that came from: image of Japanese knotweed.

The plant looks a little like bamboo, but is much easier to grow in cold places. It adds a pleasantly exotic touch to any non-Asian garden – and when deprived of its natural enemies it can flourish. It readily establishes itself in temperate climates along rivers, roadsides and waste ground. It forms thick dense colonies and rapidly crowds out competition. It can do well in many different kinds of soil: it's not fussy about pH balance or salinity. It rises from a thick system of rhizomes – and that is the problem for humans. It is not just that the plant compromises natural ecosystems. It also destroys human structures.

That is quite hard to pre-empt, because Japanese knotweed is not easy to identify, readily confused with dogwood, lilac, Himalayan balsam, bamboo and others. Once the root system is established, however, you have problems: the roots will damage concrete structures, buildings, railways, roads, flood defences, pavements and retaining walls. You can cope with it by removing everything growing above the ground, but it keeps coming back and it takes several years to kill the plant off.

You can use a herbicide that travels through the plant to reach the roots, but it's much quicker to dig the stuff up, and that is the option usually taken by developers. That gives two problems: the safe disposal of the material is a legal requirement in many places (in the UK by 1981). This is labour-intensive and not invariably successful. You can kill the root system by blocking out the light, covering the site with light-tight materials – but the stalks are sharp and can penetrate fabric, let the light in again and give the plant a second chance. You can cover the site with concrete: but it had better be good. The roots beneath will find any flaw and grow into them, causing splits and cracking. The rhizomes can stay viable for twenty years, and the plant can regenerate from a small fragment: there is no such thing as partial destruction of Japanese knotweed.

In the UK, Defra – the Department for Environment, Food and Rural Affairs – estimated that national eradication of Japanese knotweed would cost £1.56 billion. During the construction of the velodrome and aquatic centre for the London Olympic Games of 2012, the cost of the removal of Japanese knotweed was £70 million. People have been refused mortgages because of the proximity of existing Japanese knotweed to the property; those who own such properties have found themselves without buyers. The UK population is believed to have been cloned from a single female, though it has also hybridised with related species. It is now found in forty-two American states. In Vancouver the plant crossed a four-lane highway.

There are continuing experiments with the plant's natural predators; these include a leaf-spot fungus specific to the plant, and a psyllid, from the order known as *Hemiptera*, or true bugs – sucking insects. This was the first time such a project was sanctioned by the European Union; we are now familiar with the problems that come from reckless attempts to combat an invasive by introducing another: it can follow the logic of the old lady who swallowed a spider to catch the fly.

There is only one sure and certain method of dealing with all invasive species, and that is not to let them in in the first place. Once they are there and have passed the searching double-test of 10 per cent, their removal becomes a colossal undertaking. Again and again throughout this book we will come up against the issue of human control. It is incontrovertible that humans have affected just about all life on the planet. But there is a difference between affecting and controlling.

FIFTY-THREE

BAMBOO

A flash of lightning
the sound of dew
dripping down the bamboos

Buson

I remember the first time I saw it: an ultra-modern, state-of-the-art skyscraper under construction, enclosed in a giant cage of scaffolding. But the scaffolding was bamboo: an immense series of bamboo staves lashed together, reaching several hundred feet above where I stood on Connaught Road in Hong Kong. This collision of Chinese-restaurant kitsch and modern technology was bewildering, and seemed recklessly primitive. Later I did regular work for a Hong Kong-based construction magazine; I learned that many a typhoon has stripped buildings of conventional steel scaffolding while the bamboo stood firm on the building next door. Bamboo has a greater tensile strength – resistance to breaking or splitting under tension – than steel: 28,000 pounds per square inch compared to 23,000. It is much safer – falling bamboo poles are less reliably lethal – six times quicker to put up and twelve times quicker to take down. Bamboo is seen in the West as an aspect of Eastern décor, but it's one of the great plants of human civilisation.

Bamboo belongs to the family *Poaceae*, or grasses, in its own subfamily of *Bambusoideae*. It contains around 1,400 species that range in size from 6 inches (15 cm) to 100 feet (30 metres), or even taller. They can have a diameter up to a foot (30 cm). When they break the earth as shoots they are already at their full diameter, which they retain to the summit without tapering. They grow to their full height in a single growing season, and are among the world's fastest growing plants. They have been known to grow 3 feet (0.9 metres) in twenty-four hours, 1.5 inches (4 cm) in an hour, 0.3 inches (7.5 mm) in ninety seconds; such speeds depend on species, soil type and local conditions. They have hollow stems, in the manner of grasses, and the giant species are the most massive of their family. They grow to their full height before branching at the nodes: these are the junctions between sections of the stalk that give the plants their very clear identity.

Each shoot springs from a rhizome; there is a broad distinction between clumping and running species; the latter spreads faster and along straight lines. They flower rarely: most species every sixty to one hundred and thirty years. The

Many virtues: The Seven Sages of the Bamboo Grove *by artist Fu Baoshi (mid twentieth century).*

species tend to flower at the same time all over the world, and the huge stress this event places on the plant kills it. Sexual reproduction is a big deal in the bamboos: the process of cloning is its daily staple and that's how bamboos are managed under cultivation. The rhizomes in a patch of cultivated bamboo can remain productive for half a century, even with heavy exploitation.

Bamboos are found mostly in the Asian tropics, but there are species in the Americas, Australasia and Africa. They have been put to so many and such various uses as to tempt an observer to a Panglossian view of the world, and assume that bamboos were put there for the benefit of humankind. The large, woody 'timber' bamboos have been used for construction for thousands of years, harvested at the time of their greatest strength and smallest infestation from pests, when the sugar content is at its lowest. Their strength and especially their durability depends on the skill in their handling and preparation. There is a bamboo suspension bridge in China said to date from the third century BC: its survival is the result of continual maintenance and renewal.

Bamboo has been used to make weapons: as staves, as spears, as arrows. Indian police today carry a bamboo stave called a *lathi*, 2-3 feet (60-90 cm) in length. Bamboo was used in the construction of the first gunpowder-based weapon, the fire-lance; it was in use in the twelfth century and was a combination of flame-thrower and ultra-short-range shotgun. This weapon was highly effective and was the ancestor of all modern firearms. Would there be guns if there was no bamboo?

Bamboo has also been used to make musical instruments: flutes, panpipes, drums, xylophones, marimbas. It is used for baskets: the development of containers was a major advance in human civilisation (see Chapter 55 on gourds), making people more mobile and versatile. Bamboo is also used for making furniture, fish-traps and fishing poles. Bamboo rods were the choice of recreational anglers or fishers until about 1950, when fibreglass became more prevalent; they were called bamboo rods in America and split-cane rods in the UK. When I lived on Lamma Island near Hong Kong, every year a theatre of bamboo was constructed on the football pitch: every night for a fortnight Chinese opera was performed inside to an enthralled and noisily eating audience.

Bamboo can also be used to make firecrackers: you fill a short length with water and calcium carbide and – with a long stick if you are prudent – light the acetylene gas that is generated. Bamboo has been used for rafts and for floating houses. Bamboo was used to make surfaces to write on at least 3,000 years ago, and pens to do the writing can be made from bamboo.

Bamboo shoots can be eaten, though they need proper cooking: without it the toxins they contain can produce cyanide in the gut. They are often tinned and are a routine ingredient in Chinese cuisine; they add bulk and texture to stir-fries and other dishes, without contributing much in the way of flavour. Bamboo shoots can also be grated, fermented and pickled.

It is not surprising that so supremely useful and versatile a plant plays a significant part of the cultures of the people who use it. Such lore has even penetrated the West: we are most of us familiar with the proverb about the bamboo's strength. It was used by Bruce Lee, the great martial arts film star of the 1970s: 'Notice that the stiffest tree is more easily cracked while the bamboo or willow survives by bending with the wind.'

In Chinese culture the bamboo represents many important virtues: it is upright, tenacious, modest, elegant, plain and full of integrity. Its strong roots represent resoluteness, its tall straight stems honour, its hollow interior modesty, and its clean exterior chastity. It is therefore one of the Four Gentleman, along with plum blossom, orchid and chrysanthemums, all of which stand for different virtues. Bamboo is also one of the three friends of winter, along with pine and plum blossom: the pine keeps its leaves, the bamboo doesn't wither and the plum blossom appears gloriously early in the year. In the Philippines there is a creation myth in which the first man and the first woman were made from opposite halves of a split bamboo. In Malaysia there is a story of the dreamer who dreams of a beautiful woman, wakes up and discovers her hidden inside bamboo.

Bamboo has been a subject of art for centuries: in 1633 Hu Zhengyan produced *Ten Bamboo Studio Manual of Calligraphy and Painting*: in such pen-and-ink work the calligraphy and the bamboo leaves seem almost the same thing.

In recent times bamboo has been planted as a quick-growing answer to the problems of carbon sequestration. In China the ancient wild bamboo forests have been protected as homes to the giant panda, which feeds almost exclusively on bamboo. Bamboo is used to make garments, including socks and underclothes: a form of viscose rayon is made by dissolving the cellulose in bamboo and allowing it to form fibres.

You may be tempted to draw many generalisations from this chapter, and, indeed, from many others in this book. I will close with a classic Zen saying: 'From the pine tree learn of the pine; from the bamboo learn of the bamboo.'

FIFTY-FOUR

TEA

The trouble with tea is that originally
it was quite a good drink.

George Mikes

After water, the world's most frequently consumed drink is tea. We drink more tea than coffee, soft drinks and alcoholic drinks put together. Tea identifies cultures and social classes within cultures. It is a symbol of hospitality. It defines nations. It caused a revolution. It created the most notorious trade war in history. The bush has helped to make us who we are and to set the agenda for our social, commercial and political history.

We are dealing with a single species, *Camellia sinensis*, from the family *Theaceae*. Left unattended, tea plants grow into modest trees, as high as 50 feet (16 metres), but they are normally pruned to waist height for easier harvesting. According to the foundation myth, in the year 2737 BC the Emperor Shen Nong, a wise ruler and scientist, was boiling water in his garden when a leaf fell into the liquid. It was of course a tealeaf: he drank the liquid and was delighted. In another story Bodhidharma, who brought Buddhism to China and is considered the founder of Zen, attempted to make a point by meditating for nine years without sleeping. Furious at his failure, he cut off his eyelids and flung them to the earth – from which tea bushes instantly sprang up.

This second story pinpoints the essential appeal of tea: a mild caffeine hit. Roughly speaking, a cup of coffee will have about twice the caffeine as a cup of tea made from black leaves. Tea became a widespread drink in the Tang Dynasty of AD 618-917. Lu Yu wrote *The Classic of Tea* in 762, which tells you how to grow and how best to enjoy it, relating the drink to Buddhism, Taoism and Confucianism. It was essentially about harmony and simplicity – the guiding principles of the universe. To drink tea in the right way leads to deeper understanding of the great matters of life.

There are two odd side effects from the ancient cultural importance of tea to the Chinese. Drinking tea involves boiling, and is therefore a relatively safe way of drinking water: untreated water is liable to become a culture for dangerous bacteria. The Western tradition made its own safer drink by fermentation (see Chapter 11 on yeast). As a result, genetic intolerance of alcohol is rare in Western

More tea? A Conversation at Tea Time, *1927, by Frank Moss Bennett (1874-1952).*

people, relatively common among the Chinese (those afflicted go red in the face and are mildly drunk after a few mouthfuls). The consumption of alcohol tends to involve glass, because people like to see the liquid's colour in both storage and drinking vessels – bottles and glasses. Glass technology developed in the West, but hardly at all in the East. The Chinese led the way in many technologies (as we saw as recently as the last chapter on bamboo with regard to firearms), but they lagged behind in glass. The West invented telescopes and microscopes to enhance learning – and also spectacles, which extend the effective life of a learned person by a good twenty years.

Tea drinking spread to Japan and Korea. It reached Europe in the seventeenth century through Portuguese and Dutch merchants. Charles II's wife Catherine of Braganza, who was Portuguese, brought tea to England. Charles himself had acquired a taste for the drink during his exile in Holland. In a diary entry for 1660,

Samuel Pepys wrote: 'And afterwards I did send for a cup of tee (a China drink) of which I never had drank before, and went away.'

The drink gained popularity. It was seen as a more feminine and domestic drink than coffee, which men drank together in coffee houses. The seventeenth-century chronicler of the French aristocracy Madame de Sévigné wrote: 'Saw the Princesse de Tarente, who takes 12 cups of tea every day... which, she says, cures all her ills. She assured me that Monsieur de Landgrave drank 40 cups every morning. "But Madame, perhaps it is really only 30 or so." "No, 40. He was dying, and it brought him back to life before our eyes."'

The popularity of tea spread during the eighteenth century. In Russia it became customary to drink tea with lemon, the water heated in a samovar. The Dutch took tea to New Amsterdam, which the British took over and renamed New York. As the popularity of the drink spread it became a money-spinner. The British East India Company, trading directly with China, grew rich. Taxes on tea helped to protect the company. This was resented in Britain's American colonies. In 1773 a group dressed as Native Americans raided a British tea ship and poured the cargo into Boston Harbor in protest. The Boston Tea Party was the beginning of the American Revolution; the United States was established in 1776.

For many years China was the only source of tea. This led to a serious trade imbalance between China and Western powers, including Britain. Attempts to redress this by selling opium – seen as a useful medicine (see Chapter 57) – to China led to resistance from the Chinese government. The first of the Opium Wars took place in 1839-42. This and a later conflict, also involving France, were emphatically won by the West.

By this time tea-growing had begun in India. Robert Fortune, an English botanist, played an important part in this decisive shift, working in China disguised as a Chinese merchant, bringing seeds out to India, and bringing Chinese experts to India. The East India Company's monopoly was eventually broken in the 1870s. Even before this, tea clippers – very fast sailing ships – raced to bring the freshest tea back to Europe, some of them equipped with as many as thirty-five sails. Tea became cheaper and more democratic. It was drunk by working-class people in Britain at the end of a day's work, to accompany the evening meal, which was consumed early to replenish energy.

English obsessions with class express themselves readily through tea. Today, to refer to your evening meal as 'tea' is to make no bones about your working-class origins. Afternoon tea, taken at a low table with sandwiches and cakes, grew popular in Victorian times. (It also meant you could take your evening meal much later, which showed you weren't working class.) Working-class people traditionally put the milk in before pouring tea from the teapot; those with higher social pretensions put the tea in before the milk.

Tea became an important drink for pauses in the working day: the drink supplied a lift from both the caffeine and the generous spoons of sugar (see Chapter 79). A tea break became a working man's right. In the late twentieth century there was a growing tendency for the middle classes to drink more exotic teas: Earl Grey, which is flavoured with the citrus plant bergamot, or lapsang souchong, which is smoked over a fire. It grew customary to refer to the more conventional black tea as 'builder's tea'; a preference for which can be loudly expressed by people from the middle and upper classes. A joke from the 1980s: why do anarchists drink Earl Grey? Because proper tea is theft.

India, formerly just a tea-growing nation, became a tea-drinking nation; heavily sweetened tea is sold on every railway station. The country with the highest per capita tea consumption is Turkey, where they drink it from a glass. At the World Fair in St Louis in 1904 a company planned to offer hot tea to visitors, but the weather was too warm for the offer to be attractive – so they linked up with an ice cream manufacturer and made iced tea. To this day 80 per cent of all the tea drunk in the United States is iced.

Attempts to market instant tea have been unsuccessful. In the Second World War, British and Canadian troops were issued with Compo, a hard block that contained tea, milk and sugar. You mixed this with boiling water; as it cooled the

liquid grew a thick scum on its surface. Tea bags using fabric had been tried and patented in the early twentieth century, but only grew popular in the 1950s when paper was used instead. Tetley's introduced the tea bag to Britain, where it was an instant success. What was lost from the ritual of the teapot was compensated for in convenience.

Tea remains important. In Britain it is a symbol of hospitality: you offer a cup of tea to a visitor as a matter of course. People who visit your house for professional reasons, whether to help with the accounts or the pipes, will be offered tea. Coffee is a rival (and a fuzzy indicator of social class) but offering tea is important.

Tea is drunk with mint and sugar in North Africa, with yak butter in Tibet, with jasmine flowers in China. The reasons for adding milk are warmly debated among those who care: adding milk first cooled the tea and made it less likely that delicate porcelain cups and bowls would crack; tea arriving from a long sea voyage was bitter and needed to be mellowed; milk cut the oxalic acid it contained. My mother used to tell a story about an Oxford professor receiving a Chinese student in his rooms, serving tea in the traditional British style. 'Nothing like tea,' the professor said self-approvingly. His student agreed politely: 'No, it isn't.'

A landscape of tea: tea-gathering on Mr Holles's Plantation at Garoet, Java; beyond the volcano Goentoen.

FIFTY-FIVE

GOURD

'I'm very glad,' said Pooh happily, 'that I thought
of giving you a Useful Pot to put things in.'

A. A. Milne, *Winnie-the-Pooh*

What was the big breakthrough? What was the most significant invention of
our hunter-gathering ancestors? What was the real game-changer, the one
that set these ground-apes on course for their descendants' domination of the
planet? There are two conventional answers to this. We have already discussed
the mastery of fire (Chapter 22, kigelia); the use of stone and bone tools is beyond
the scope of this book.

The above questions are impossible to answer, because we weren't there in any
but the genetic sense, and because over the course of uncountable millennia a
great deal of evidence gets destroyed. Almost every trace of what our most distant
ancestors did with plants has returned to the economy of the soil. It's possible that
we over-value stone tools – we talk about the Palaeolithic and Neolithic periods,
Old and New Stone Age – because the evidence of other and perhaps more
important inventions has been lost. Two points should be made here. The first is
that the shortage of evidence opens the way for speculation, which should of course
be sober and sensible; the second is that there exists no hierarchy of importance of
inventions, any more than there is a hierarchy of the best pop songs. All
breakthroughs have their significance; though it's possible that some have been
underestimated or even lost because of lack of evidence.

We humans live our lives with things you can put things in. I am working in winter
as I write these words, and a count reveals that I have access to thirteen pockets,
seven of which are in use. There's a backpack in the corner ready for my next journey.
In the kitchen the food and the drink is ready for use in packets and jars and bottles.
But so far as I have been able to establish, no non-human species has ever used a
container. The use of tools was once said to define humans, though over the past
half-century it's been established that plenty of species use tools, including octopuses.

But putting things in things may be a wholly human idea. So what started it? In
Africa to this day, gourds and calabashes are used to store and transport liquids

Opposite Portable drinkables: Sancho Panza, 1859, by John Gilbert (1817-97), pictured with useful gourds.

and loose solids like grains, and this is likely to be a tradition that goes back for many thousands of years: long before the invention of agriculture – which wouldn't have been possible without containers.

The most useful gourds come from the family *Cucurbitaceae*, which is found all over the world in tropical and warm temperate climates. There are getting on for 1,000 species of mostly annual vines that tend to grow large fruit: modern cultivated examples include cucumbers, melons, marrows, courgettes (zucchini), pumpkins and squash. The classic bottle gourd *Lagenaria siceraria*, which has its origin in Africa, comes from the same family. It's nice to imagine some ancient (Palaeolithic) Archimedes finding such a gourd – fallen from the vine, hollowed out by rodents and ants, baked hard in the sun – dipping it into a stream and drinking deep, before toasting the horizons with a cry of eureka.

Containers change the possibilities of life. With a gourd full of water you can travel further, especially if you can attach the gourd to your person and leave your hands free (see Chapter 12 on hemp). If you have a base, however temporary, you can keep water there and reduce the number of trips you make to the river, where the crocodiles live. You can put food in a container and keep it safe from rodents and insects, even more so once you have mastered the technology of the stopper. The gourds may not be perfectly flat at the bottom but they stand with perfect security if you place them in a ring of dried grass or, still easier, use three pebbles – 'African physics' as my botanist friend Manny Mvula says. Containers radically change your array of options. Gourds have been used wherever it's warm enough to grow them, but the invention can only have started in Africa – where they made possible the spread of humanity from Africa across the world.

From there it is a short step to improving the technology of gourds: first by hollowing them out on purpose, perhaps with wooden or stone tools. Some species have an edible interior, especially if harvested when young, but a more mature or otherwise inedible specimen should be left to dry in the sun before it is hollowed out. There are two ways of improving the watertightness of the container. The first is by heat: you put hot ashes and coals into the interior and swirl them about. This takes nice judgement: in unskilled hands a fine container can be spoiled by overheating. Others prefer to use beeswax, which must first be heated to make it soft and pliable. You run this over the inside; with a wide-mouth gourd you can use your hands, but with a narrow-mouth, use feathers on a stick. The smaller bottle gourds have an additional use: you can turn them sideways and remove what is now the top, making for yourself a long-handled drinking vessel.

Though the terms gourd and calabash are bandied out indiscriminately, technically a calabash is the fruit of a calabash tree, *Crescentia cujete*, which is found in Africa and Central and South America. These fruit also make excellent containers, and are managed in much the same way.

Once gourds became part of human technology, many more uses for them were found. They have been used as household and personal ornaments. They have been used to make musical instruments: drums, rattles, stringed instruments, and even nose-flutes. They are still used to make sophisticated modern instruments; my son, a musician, owns and plays a sitar that, like all sitars, is made from a tumba gourd; this is what gives the instrument its distinctive resonance. You can use the interior of the Luffa species to scrub yourself in the bath. Every year in the United States and elsewhere, hollowed-out pumpkins are used to chase away evil spirits on Halloween.

What would human civilisation have been like if a species of African vine hadn't happened to grow in a shape of what we now recognise as a bottle? Perhaps we should all raise a glass to that Palaeolithic Archimedes – and more especially to the forces of life that create such extraordinary stuff.

FIFTY-SIX
VIRGINIA CREEPER

I procured various other kinds of climbing plants,
and studied the whole subject.

Charles Darwin

The margins, the extremes, no man's land, Tom Tiddler's ground – those dangerous areas where you find variations and exceptions, monstrosities and absurdities – can be powerfully instructive. Here among the outliers you can find some obvious truths: and sometimes they turn out to have a general application. It's no wonder Charles Darwin felt at home there.

Darwin, as we have seen (Chapters 13, 27, 33 and 44), was always re-proving his theory of evolution by means of natural selection. This was partly because until it won general acceptance – even in the scientific community this was a much slower process than we like to imagine – it could never be proved often enough. It was also because if you investigate any form of life, you can't help but come up against the questions of why and how, and the answers lay then and now in the principles laid out in *On the Origin of Species*. Everything he wrote before was a prequel; everything that came after was a sequel, piling demonstration on demonstration, proof on proof.

I don't think Darwin was consciously seeking such repeated vindication. It was just that he got sucked in. If something aroused his interest it was likely to become the centre of his life for the next five years. Darwin was like a fulfilled Toad, the character in Kenneth Grahame's *The Wind in the Willows*: he would take up his latest craze and it would become the most important thing in the world. Unlike Toad, he would stick at it until he had cracked it: until he had learned all he could and then imparted his new-found knowledge.

So when Darwin was sent a few vine seeds from the *Cucurbitaceae* family (see previous chapter on gourds) and his attention was caught, that was it, for quite a long time. He received the seeds from the American botanist Asa Grey in 1858, along with a short paper on the subject. In 1865, Darwin published *On the Movements and Habits of Climbing Plants*.

Climbers throw up a number of odd questions. They are unlike conventional plants that support themselves – but if you can show that they abide by the same rules as other plants in order to lead their very different lives, you've established something important – common ancestry. Climbing plants appear to think and to

A plant to make you think: Virginia creeper growing on Down House, home of Charles Darwin, painted in 1880 by Albert Goodwin (1845-1932).

move towards a clear and obvious goal. How can they do this without violating the parsimonious principles of the *Origin*? It's also true that the *Origin* concentrates on the animal kingdom: if Darwin could demonstrate that the same principles operate for plants he would be pushing his theory far beyond the possibilities of denial.

Why climb at all? 'Plants become climbers, in order, it is presumed, to reach the light,' Darwin wrote. All plants grow towards the light, but climbers take a short cut, using other plants and other natural (or human-made) structures for support, rather than growing rigid stems. 'This is effected by climbers with wonderfully little expenditure of organised matter, in comparison with trees.' We have looked at rainforest giants, like the Brazil nut tree shooting 160 feet (50 metres) high (Chapter 14): a climber can reach the same height without troubling to grow a woody support system is a classic example of the opportunistic nature of evolution.

Climbing plants do not make up a single group. They are a bunch of often unrelated plants that operate on the same basic idea. They all reach light that would otherwise be inaccessible: but they do so in a number of different ways. They have reached the same solution to the same problem by different means: what's known as convergent evolution. (Classic example: bats, insects, birds and the extinct pterosaurs all evolved true flight; each one by a different evolutionary route.)

Darwin was enthralled by these plants. 'Some of the adaptations displayed by climbing plants are as beautiful as those by orchids for ensuring cross-fertilisation,' he said, lofty praise indeed. Darwin loved conceptual beauty: the revelation of perfect simplicity behind extraordinary complexity. He had a leap of understanding about the way that twining plants wind themselves around a support. They don't do so by sense of touch, he suggested: they do so in the manner of a man swinging a rope around his head and side-swiping a stick. It's a classic example of the simplest, most economical explanation.

The Virginia creeper is a native to North America, and was introduced to Europe (another part of the Columbian exchange) as an ornamental plant. It will climb up the side of a house without damaging the masonry; it doesn't grow into cracks but supports itself on self-grown stickers. He wrote:

> The gain in strength and durability in a tendril after its first attachment is something wonderful. There are tendrils now attached to my house which are strong and have been exposed to the weather in a dead state for 14 or 15 years. One single lateral branchlet of a tendril estimated to be 10 years old, was still elastic and supported a weight of exactly two pounds [0.9 kg]... so that for having been exposed during 10 years to the weather it would probably have resisted a strain of ten pounds [4.5 kg]!

Tendrils and suckers are more specialised apparatus than mere twining stems. They are like leaves, but different: leaves that have been modified: leaves that have changed their function. They were not designed from scratch as tendrils but were adapted from pre-existing structures. That is another essential principle of evolution. Classic example: some dinosaurs evolved feathers for thermoregulation: feathers turned out to be immensely suitable for flight. Another: our mammalian lungs and the swim bladders of ray-finned fishes are essentially the same organ, differently adapted. This principle was formerly known as preadaptation, but this term has been rejected because it seemed to imply a plan, a goal, a teleological system. Since 1982 the term exaptation, coined by Stephen Jay Gould and Elisabeth Vrba, is preferred. The suckers on a Virginia creeper demonstrate this principle perfectly.

Darwin's investigations showed that plants don't just grow: they also move. The tendrils and growing shoots of climbing plants revolve in search of something to hold on to: and some tendrils are, indeed, capable of noticing when they touch an object (Darwin used the term 'irritability'). Movement, a function of hydraulics rather than muscles, is essential for the life of the climbers. Darwin wrote: 'It has often been vaguely asserted that plants are distinguished from animals by not having the power of movement. It should rather be said that plants acquire and display this power only when it is of some advantage to them.'

There in a phrase – in a plant – we find the uncompromising functionalism of evolution. There is beauty in this view of life, a beauty strong enough to ravish the mind of one of the greatest thinkers in history. Darwin's former residence, Down House in Kent, is now in the hands of English Heritage. They have planted a Virginia creeper and it grows up one of the walls: supporting not the house but the theory of evolution by means of natural selection.

FIFTY-SEVEN

OPIUM POPPY

Thou hast the keys of Paradise, oh just, subtle and mighty opium!

Thomas de Quincey, *Confessions of an English Opium-Eater*

For most of human history the opium poppy was considered a boon to humankind. Spreading its most significant product was a noble act as well as a profitable one. The Sumerians, the first known users 6,000 years ago, called it the joy plant; the physician and alchemist Paracelsus created a tincture of opium and called it laudanum, 'that which is to be praised'. Opium has inspired great art.

Most likely the plant first grew wild in the eastern Mediterranean, but as its properties became known it was widely cultivated, traded and distributed. It is one of seventy species in the *Papaver* genus, along with the field poppy (see Chapter 19). The difference is that when the immature seedpod of *Papaver somniferum* is wounded with a sharp implement, it exudes a latex packed with chemicals. These have a powerful effect on the human body. The latex relieves pain as no other substance can.

As always, no one knows how this effect was discovered, but once the knowledge got about, the plant became important to one civilisation after another: those of Assyria, Babylon and Egypt, where its cultivation centred on Oedipus's city of Thebes. The Phoenicians and the Minoans traded it around the Mediterranean. Hippocrates, though sceptical of claims for the plant's magical properties, acknowledged its usefulness as a narcotic and as a treatment for female disorders. It was administered on a sponge to patients undergoing surgery. Alexander the Great brought it into Persia and India, where it flourished. Poppies can be found decorating Greek statues of gods. Pliny the Elder praised its power to give sleep and gives a handy test for its purity: if it burns with a bright clear flame it's pure, but if it's hard to light and keeps going out, it's adulterated. By AD 400 Arab traders had introduced it to China.

Opium use flourished in Asia but declined in Europe until it was reintroduced by the Portuguese around 1500, when people started smoking it. This method affects the body very rapidly and makes it an unambiguous drug of pleasure; it was by then a recreational drug in Persia and India. At the same time, the sixteenth-century Swiss physician and alchemist Theophrastus von Hohenheim (or Paracelsus) developed the medical use of opium, creating the substance he called laudanum.

This is principally opium in ethanol, ready for immediate use. It was an important part of the so-called Medical Revolution in which observation was considered at least as important as received wisdom.

The desirability of opium was clear to all and was increasingly traded. The English apothecary Thomas Sydenham developed Sydenham's Laudanum, which mixed opium with sherry and herbs. The Dutch traded opium from India to the Chinese; the British East India Company took control of the opium-growing areas of Bengal and Bihar and pursued the trade with China on their own behalf; by the end of the eighteenth century they had established a monopoly. Indian farmers were forced to grow opium by the established methods of threat, violence and debt.

The British Empire attempted to redress a massive trade imbalance with China (partly created by the craze for tea, see Chapter 54) by selling them opium. True, the nature of the drug was not properly understood and the concept of addiction was a later development, but this is not an edifying tale. It should be added that Britain also imported a great deal of opium for its own use: in 1830, a total of 22,000 pounds (10,000kg). The Chinese attempted to ban the import of opium and restrict its use to medical purposes. Foreign traders were ordered to surrender their opium. The British sent in the warships and the first Opium War began in 1839; Britain defeated the Chinese in 1841. The Chinese handed over Hong Kong in reparation; it remained in British hands until 1997.

In 1803 Friedrich Sertürner of Germany discovered the active ingredient in opium. He dissolved opium in acid and then neutralised it with ammonia and this gave him what he called Principium somniferum – or morphine. Opium had been tamed; more, it had been perfected. It was considered God's own medicine.

Opium was celebrated in the Romantic Movement as a portal to enlightenment. In 1821, Thomas de Quincey published *Confessions of an English Opium-Eater*, an early chronicle of addiction. Perhaps the most famous user was Samuel Taylor Coleridge; his poem *Kubla Khan* is a celebrated work of ecstatic visions:

Weave a circle round him thrice,
And close your eyes with holy dread
For he on honey-dew hath fed
And drunk the milk of Paradise

Hector Berlioz wrote his *Symphonie Fantastique* in 1830, scored for more than ninety instruments. In what the twentieth-century musician Leonard Bernstein called the first piece of psychedelic music, the symphony tells the story of the unrequited love of an artist who poisons himself with opium and dreams of his own

Opposite Bewitching: Beata Beatrix, 1864-82, by Dante Gabriel Rossetti (1828-82); beside her a glowing opium poppy.

execution. The importance of opium to the Romantic Movement has been much celebrated; the critic Elizabeth Schneider issued a corrective: 'It does not give him powers that he did not have or change the character of his normal powers.' In other words, you can't take a drug and become a great poet, though the drug might well affect the stuff a great poet produces.

To the modern eye the truly striking part of the business is that opium was available to anyone who could afford it. You could just go to a shop and buy it: headache, period pains and writer's block could be treated with over-the-counter opium. Coleridge used up to 100 drops of laudanum in a day. The United States president William Henry Harrison was treated with opium in 1841. The Union army used 175,000 pounds (80,000kg) of opium during the American Civil War.

In 1874 the English chemist Alder Wright synthesised heroin from morphine. This breakthrough was celebrated as a non-addictive substitute for morphine. Morphine is treated with acetic anhydride and then purified to make heroin. It was marketed by Bayer as a children's cough medicine.

But these carefree days were coming to an end. In 1906 the United States brought in the Pure Food and Drug Act that required clear labelling. The USA banned the import of opium three years later. That same year, 1909, the International Opium Commission met in Shanghai, and the dangerous and damaging nature of the drug was no longer denied. The following year the British were finally persuaded to stop trading opium. In the United States, President Theodore Roosevelt appointed Dr Hamilton Wright as Opium Commissioner. In 1911, Wright said: 'Of all the nations of the world, the United States consumes most habit-forming drugs per capita. Opium, the most pernicious drug known to humanity, is surrounded, in this country, with far fewer safeguards than any other nation in Europe fences it with.' The Harrison Narcotics Act was passed in 1914.

The criminalisation of opium and its derivatives inevitably created an illegal trade. It helped that morphine and heroin are much less bulky than raw opium and therefore easier to smuggle. The Second World War did a great deal to restrict the trade, but it rose again after the war. The United States saw the spread of Communism as the world's most important issue and through the Central Intelligence Agency they established alliances with potential anti-Communist leaders in many places, including the land between Cambodia, Laos and Burma, the so-called Golden Triangle.

The supply of weapons and transport by the USA and France enabled the warlords to establish a highly profitable illegal trade in opium. This surge in production forced prices down and made widespread addiction possible. William Burroughs wrote about addiction in his memoir *Junkie* and in his powerful novel *Naked Lunch*. The Velvet Underground included the frightening song 'Heroin' on their 1967 album; John Lennon released his song 'Cold Turkey', about heroin withdrawal, in 1969. The use of heroin had acquired a dreadful glamour: the privileged addict who lives in a world no non-addict can ever understand.

There are many legitimate medical uses of opioids, the synthetic and semi-synthetic drugs that operate like opium. These include codeine. More powerful opioids are used to treat acute pain; they are also used in palliative care of people suffering from cancer and rheumatoid arthritis.

The United Nations Office on Drugs and Crime estimates that 15 million people worldwide use opioids illicitly, the majority using heroin. The trade is worth an estimated US$55 billion annually. The main source these days is Afghanistan, followed by Myanmar, Mexico and Colombia. The trade is linked not just with straightforward criminals, but also as a funding operation for terrorist groups.

The plant *Papaver somniferum* is also cultivated as an attractive garden plant. The seeds can be used in cooking: very pleasant when sprinkled on bread rolls.

FIFTY-EIGHT
BANANA

Yes! we have no bananas,
We have no bananas today.

Frank Silver and Irving Cohn

There is something reassuringly comic about bananas: their phallic shape, their extravagant colour, the pleasant blandness of the fruit inside, and above all, the reputation of the skin as something to slip on and make the world laugh. 'Sliding' Billy Watson, an American vaudeville artist of the early twentieth century, is credited with the first banana pratfall, Charlie Chaplin with the first banana-skin tumble on film. Bananas have become a standard racial slur; the great Japanese poet Basho named himself after a banana tree; and there is a terrible fear that an event of the 1950s known as the banana apocalypse could be repeated.

We usually use the word banana to refer to the fruit, or, to be more accurate, the berry. The banana plant is not strictly a tree but a herbaceous plant; it has no woody stem, and its aerial parts die back to the ground at the end of each growing season, which in most cultivated bananas is a little over a year... but the plant sprouts again from its rhizome beneath the surface for another half-dozen growing seasons.

Evidence of banana cultivation has been found in the Kuk Swamp in Papua New Guinea dating back 10,000 years. The plant spread into Southeast Asia. It grows well in deep, loose soil in the moist tropics. Bananas were taken up by Arab traders and spread with Islam; they were later traded by the Portuguese. Bananas are not a recent arrival in Europe, as is often assumed: Alexander the Great noted them in 327 BC on his way to India. Pliny the Elder wrote about them: 'The fruit has a wonderfully sweet juice and a hand provides enough fruit for four people.' In the twelfth century Ibn al-'Awwam wrote his great *Book of Agriculture*, in which he described the art of banana cultivation; the fruit has a strong Islamic tradition as a fast-breaker during Ramadan. Few food items are as effective at delivering rapid nourishment, which is why cyclists on the Tour de France and tennis players in long matches routinely refuel with bananas.

Most readers will be familiar with the sweet or dessert banana: the default banana. But in many part of the world bananas are a carbohydrate-rich main-course staple. These are sometimes referred to as plantains, but they're the same thing... though that's being a little approximate. Banana plants come from the

HOLESOME, NUTRITIOUS, DELICIOUS TO EAT
TROPICAL TREAT THAT'S HARD TO BEAT ···

Irresistible: American poster advertising tropical fruits from South America, c.1930.

Musa genus, as defined by Linnaeus. He succeeded in growing a plant to fruiting and named the species *Musa paradisiaca*, a reference to the belief that what Eve offered Adam in the Garden of Eden was not an apple but a banana. Two species are now recognised as the basis of all commercial plants: M. *acaminata* and M. *balbisiana* – but they have been hybridised again and again in the creation of different cultivars and are now pretty well indistinguishable. To travel in, say, Peninsular Malaysia is to understand the paucity of the West's banana universe. We restrict our banana ideas to a single variety, the ubiquitous Cavendish: Malaysian markets offer bananas of green, yellow, red, purple and brown, of different sizes, textures, flavours and degrees of sweetness.

Bananas are ridiculously easy to eat. Within the flesh there are no seeds: only, at best, tiny black specks that are vestigial seeds. Wild bananas produce fruit full

241

of seeds so that they can reproduce themselves sexually, but cultivated varieties have had the seeds bred out of them. They are reproduced by cloning, from the rhizomes. The plants grow up on their false trunks (pseudostems) and produce their giant leaves (Pliny compared them to birds' wings) and then produce a flower spike. The fruit develop from this: fifty to one hundred and fifty per plant, in bunches or hands of ten to twenty. After fruiting the plant dies back and is chopped down. From first planting it takes around fifteen months for the plants to become productive; after that harvesting is more or less continuous. The plant will produce new shoots from the rhizomes, which will in turn bear fruit.

It follows that all the Cavendish bananas we eat are more or less genetically identical: and therefore suffer from the lack of resilience inherent in every monoculture. The most popular dessert banana of the early twentieth century was the Gros Michel. In the 1950s they suffered from a fungal infection known as Panama Disease. Because of the lack of diversity there could be no resilient outliers and their cultivation became impossible. They are said to have had a richer flavour than the Cavendish, but they're gone for good: victims of the Banana Apocalypse. The same fate could overtake the Cavendish.

The problems of keeping tropical fruit fresh meant that for years bananas were an exoticism in northerly countries. Jules Verne described them in tones of wonder in *Around the World in 80 Days*, published in 1872: 'They stopped under a clump of bananas, the fruit of which, as healthy as bread and as succulent as cream, was amply partaken of and appreciated.' In 1876 at the Centennial International Exposition in Philadelphia, you could buy a banana wrapped in foil for 10 cents: a great novelty. But by 1879, *Harper's Weekly* wrote: 'Whosoever throws banana skins on the sidewalk does a great unkindness to the public.' In 1909, St Louis made throwing banana skins in the street illegal.

The exotic nature of bananas has always been part of their charm. An admirer planted a banana tree or *basho* at the hut of a great eighteenth-century poet. He took on the name Basho in appreciation and wrote the haiku:

Banana leaves hanging
round my hut
must be moon-viewing

Bananas reach their consumers in colder climates by way of the ripening room. They are harvested unripe; in their destination countries they are kept in airtight rooms and treated with ethylene gas, which creates the vivid yellow colour; you can buy ungassed fruit from health food shops. Bananas are one of the world's most important fruits (using the term informally or commercially rather than botanically); also one of the most wasteful, because the entire plant is discarded when the fruit have been harvested.

Bananas are an important crop in many tropical countries. This has given rise to the term 'banana republic', a derisive term for an unstable nation with an autocratic and possibly delusional president. There is a picturesque notion that monkeys subsist on bananas. This has given rise to a racial insult: bigoted football supporters throw bananas at black players. In 2014 the Barcelona player Dani Alves responded by peeling and eating the banana; from his expression it was clear that he was fuelling his energy for a bitter struggle. To call someone a banana is an insult in some Asian countries, suggesting that a person with Asian features has rejected the native culture and so is yellow on the outside, white inside. In 1923 Frank Silver and Irving Cohn published their song 'Yes! We Have No Bananas'; it was for decades the best-selling sheet music in history. In 1955 Harry Belafonte, the Jamaican American singer, released 'Day-O (Banana Boat Song)', based on a Jamaican calypso about dock-workers on night-shift loading a banana-boat:

Stack bananas till the morning come!
(Daylight come and we want to go home)

The Velvet Underground album mentioned in the previous chapter – *The Velvet Underground & Nico* – had on its cover a banana image designed by Andy Warhol: in early examples the banana could be peeled off to reveal a pink fruit. A good copy with an unpeeled banana will sell for around four figures – pounds or US dollars.

FIFTY-NINE

ASPIDISTRA

'You're dishonoured somehow. You've sinned. Sinned against the aspidistra.'

George Orwell, *Keep the Aspidistra Flying*

We build houses to protect ourselves from nature: once we have done so we invite nature back in. Nothing expresses the paradoxical nature of our relationship with the natural world better than the house plant. We build a wall to keep nature out, and then we go to considerable trouble to have nature inside with us. We take comfort in both the walls and the plants. We hate and fear nature: we love and need nature. To have nature inside our homes, on our own terms, tells us that we are truly civilised: that we have these opposed sides of our nature in balance. For that reason I have chosen the aspidistra to represent all house plants: the shade-loving Asian shrub that became a symbol, both loved and hated, of achievement, of comfort, of respectability.

The Sumerians and the Ancient Egyptians grew plants in decorative containers: plants for pleasure, to please the senses, to lessen the difference between indoors and outdoors. It's not clear whether they were for courtyards or for rooms inside with a roof, but in warm climates there is little difference between the two when it's not raining. Plants soften the hard lines and hard facts about the human domination of nature. Even in modern comfort we like to stay in touch with our deep past.

A house can offer a plant certain advantages, like warmth and shelter. But there are also disadvantages. The principal one is light: a structure designed to protect humans necessarily excludes a great deal of natural light. The second is water: no access to rainfall and a container with limited storage capacity. Humans can easily supply the water: light is harder, which is why shade-tolerant plants are the right ones for a house.

The Ancient Greeks and Romans also grew plants in containers, with a special taste for laurels, which are evergreen and so look good all year round. There is an additional advantage: their aromatic leaves (bay leaves) are good in cooking. The Chinese grew plants in containers 2,500 years ago. In medieval Europe monks grew medicinal plants beneath the roofs of their monasteries.

With the Renaissance and the beginning of the Age of Discovery, new plants from overseas became available. Exotic plants that would not survive outdoors could, with a certain amount of cherishing, be cultivated in buildings. Plants from West Asia and the East Indies were grown in houses in Italy, Holland and Belgium. A craze for exotic plants for the home reached England and France in the sixteenth century. An indoor plant is a statement: they can't be grown without spare income and spare labour. They are a luxury: they make it clear that their owner has risen above the struggle. But they have benefits beyond self-esteem: for many people plants make a home more agreeable, more restful, more welcoming, more homely.

In 1660 Sir Hugh Platt published *The Garden of Eden: How to Grow Plants in Homes*. This work expanded and to a considerable extent democratised the taste for house plants. Plants from South America, Asia and Africa became available by the seventeenth century, and Australasia a century later: many nurseries were established in England in the eighteenth century, selling plants suitable for the home. These could be grown in enclosed glass boxes known as Wardian cases. By the nineteenth century they were grown in the most unpretentious middle-class houses.

But a Victorian parlour was a forbidding environment for a plant. Windows were of necessity small, because glass was expensive and its technology relatively primitive, so there was limited light available. In winter, houses were heated by coal fires and the rooms were filled with the resulting pollutants (humans also suffered from related problems). Further atmospheric problems came from artificial lighting: oil lamps and coal-gas lighting also pushed out pollutants. A plant that could survive such intimidating conditions had to be tough.

The plants that did the job were the kentia palm from Lord Howe Island in Australia, the parlour palm from Mexico and Central America – and the aspidistra. This plant from China and Vietnam is a classic shade-lover. Its leaves start at the ground, without any visible stalk, as do the flowers. Around 100 species are now recognised, but *Aspidistra elatior* is the one that graced Victorian and Edwardian homes.

It became a symbol of stuffy respectability, of keeping up appearances, of what-will-the-neighbours-say gentility, a world stifling and repressive. This aspect was celebrated in George Orwell's 1936 novel *Keep the Aspidistra Flying*, in which the hero, Gordon Comstock, wages a personal war against 'the money god'. Of course, after a long and terrible struggle he loses; his only victory is his refusal to have an aspidistra in the house. Indoor plants became unfashionable: not the sort of thing that go-ahead people went in for. As a sort of counter-blast to this, Gracie Fields, the English singer, recorded in 1938 'The Biggest Aspidistra in the World':

> For years we 'ad an aspidistra in a flower-pot,
> On the whatnot, near the 'atstand in the 'all…

This celebration of ordinary folk became something of an anthem during the Second World War: a perverse symbol of pride mixed with complex layers of

irony, in the British tradition. As a result, Aspidistra was the codename chosen for a mighty wartime transmitter used by the British for propaganda and deception.

New processes for glass-making made houses lighter, and more suitable for a greater range of plants. Indoor plants came into fashion again, and by the 1970s some houses contained indoor jungles, mostly furnished with plants from tropical rainforest. This makes sense: such plants are used to a steady environment rather than violent seasonal changes. They require heavy watering and a decent level of humidity. Indoor plants have gone in and out of fashion ever since: one day I shall hang a framed machete in our sitting room with the legend 'in emergency break glass'.

House plants are good for you. A NASA study of 1989 showed that a single plant cleans the air inside a room, reducing the presence of carcinogenic substances that include benzene, formaldehyde and trichloroethylene. Well-being and productivity in schools and offices are both increased by the presence of houseplants: as a result, an industry supplies and maintains indoor plants in such places. There are less readily computable benefits from living with plants, and they are to do with biophilia. This is a term coined by the great American scientist and writer Edward O. Wilson, and it is about the human affinity for non-human life, something that everyone who has patted a dog, smelled a rose – or, for that matter, bought this book – knows all about.

The downside is that compost used for indoor plants often contains peat. Peat extraction is devastating environments and is a contributor to climate change. Non-peat substitutes exist and are used by people and firms that wish to behave ethically.

Opposite Indestructible: The Aspidistra, c.1912, by Jean Marchand (1883-1940).

SIXTY

BEAN

'What do you run on, Rockette Morton?'
'I run on beans. I run on laser beans.'

Dialogue between songs on the album *Trout Mask
Replica* by Captain Beefheart and his Magic Band

We live by consuming the energy of the sun. The most efficient and economical way of doing this is by eating plants. Another method is to consume the animals that consume plants: eating your plants at second-hand. This is more expensive of resources like land and time. In earlier times meat was an occasional luxury for most people, rather than a daily right. In the developed world most people now have the opportunity of eating meat every day – two or even three times over. This is a recent development; before that such a diet was only for the elite. The rest got their protein from elsewhere – where and when they could. Beans were known as the meat of the poor: and it's possible that Europe was changed forever with the arrival of a bean species from the New World.

Second helping? The Bean Eater, c.1580-90, by Annibale Carracci (1560-1609).

The cultivation of legumes has been part of human life since farming began. Legumes come from the family *Fabaceae*, which has many species with fat, protein-rich seeds as well as fodder crops like clover and alfalfa, with seeds unsuitable for human consumption. Legumes include peas (see Chapter 4). There is evidence of cultivation of legumes from Thailand almost 12,000 years ago, right at the beginning of agriculture; in Mexico 9,000 years ago; and in West Asia only slightly more recently. Legumes have been found in the Egyptian tombs dating back 5,000 years.

Many cultures have a tradition of combining grains with legumes: dal and rice in India, bean curd and rice in China, lima beans and corn in Mexico, chickpeas and couscous in North Africa. Legumes have a great virtue in agriculture: they take nitrogen from the air and fix it in the soil, to the great benefit of other crops. The pre-Columbian North American tradition of agriculture involves what are called the Three Sisters: beans, squash and maize (corn); the beans climb up the maize stalks and fix nitrogen for the other two.

The most important bean species today is *Phaseolus vulgaris*, which originated in the Americas. When the plant was brought to Europe it added greatly to the general availability of protein. It's been speculated that this plant, a bountiful supplier of nutrition, allowed the population of Europe to expand during the Renaissance; certainly the population of England doubled in the sixteenth century. Many different factors govern population size, so it's as well not to be too precipitate here: let us just note that the greater your access to good protein the longer you are likely to live and the more living progeny you are likely to leave.

Beans can be consumed fresh and, if young enough, their immature seeds and pods can be eaten together. But the great virtue of beans is that they are protein that keeps. They keep much better than meat, especially in a hot climate. Dried beans are good for several years, if kept safe from weevils and rodents, though they lose something of their flavour and their protein content. They can be made into good food with prolonged boiling, which is essential because undercooked beans contain toxins.

Many different bean cultivars have been produced, resulting in many different kinds of beans. Beans from *P. vulgaris* include flageolet, kidney, pinto, haricot, borlotti, cannellini, marrow and navy. Broad beans (fava beans in America) and runner beans are different species. The fibrous pods of the latter are considered palatable by some. French beans, a cultivar of *P. vulgaris*, produce an edible pod without the long fibres. Runners are sometimes called string beans.

In America beans are associated with the Old West: the idea that cowboys lived entirely on beans is deeply entrenched in global mythology, and has a fair amount of truth in it: good, portable, cheap, lightweight, long-lasting protein. It is also a fact that the consumption of beans produces flatulence. Beans contain polysaccharides that humans can't digest; these are consumed by gut bacteria in a process that creates gas, which is passed out of the human at either end.

Beans, beans, they're good for the heart!
The more you eat the more you fart!

As an old rhyme has it. The Mel Brooks film of 1974, *Blazing Saddles*, includes the famous scene of the cowboys eating their evening meal to a symphony of farts. 'How about some more beans, Mr Taggart?' 'I'd say you've had enough.'

One of the pleasures of cooking with beans is their ability to absorb flavours: garlic and herbs in Tuscan bean soup, chilli in many Mexican dishes. Beans also have a great affinity for tomatoes, another part of the Columbian exchange. They are cooked together in traditional Native American cooking, often with bear fat, maple syrup and venison. This idea was taken on and adapted by early settlers in New England; you could keep the dish warm over Saturday night and eat it for breakfast on the Sabbath, when the labours of cooking were forbidden.

This important and ubiquitous dish was an obvious choice as a convenience food when such things became possible. In 1886 the J. Heinz Company of Pennsylvania sold tins of baked beans to Fortnum & Mason in London as a luxury item, but by 1895 they had established it as a mass-market product; it was exported to all of Britain by 1901. (A tin-opener had been patented in England in 1855.) The British took to baked beans with enthusiasm: today 2.3 million British people eat them every day, mostly Heinz, which has 70 per cent of the market share, for, as an advertising slogan of 1967 had it, 'Beanz Meanz Heinz'. This is despite the fact that Heinz beans have often done poorly against their competitors in blind tastings. The dish offered is navy beans cooked in a sweetened tomato sauce. Heinz beans in the United States are made to a different recipe, one that contains 14 grams of sugar in each tin, rather than the UK's 7 grams.

In 1999 Larry Proctor, a plant-breeder from the United States, patented a variety of yellow bean (yet again from *P. vulgaris*) which he named for his wife, Enola. This gave him sole rights to sell the bean in America. He was therefore entitled to a royalty on any genetically identical beans imported into the States: and since the bean had originally come from Mexico, this was a serious problem for Mexican growers and exporters. It became a cause célèbre. It was discovered that Enola beans were part of the Andean gene pool and identical to a group of Mexican yellow beans. The patent was finally revoked in 2009 and remains a classic illustration of the need for developing nations to protect their genetic resources against what has been termed biopiracy.

Beans have always been valuable to humans. As the global population continues to grow, they will perhaps become even more important, as the ethical and economic viability of meat-eating comes more and more into question.

SIXTY-ONE
TOADSTOOL

We don't care to eat toadstools that think they are truffles.

Mark Twain

The human passion for a binary view of life is all-pervasive: and so it follows that in folk taxonomy we divide the kingdom of fungi into two: those we can eat and those that can kill us. We call the good guys mushrooms and the bad guys toadstools. But just as villains are often the most interesting characters in stories – Milton's Satan or Ian Fleming's Blofeld – so we have a fascination with poisonous fungi.

So far as a classifying mycologist is concerned, there is no practical distinction between dangerous and non-dangerous species: Caesar's mushroom, prized for its excellence as a food, is in the same genus as the death cap, the most reliably deadly of all fungi. Toadstool, the catch-all term for poisonous fungi, is not used by scientists. The enduring fascination of potentially lethal fungi is shown by the melodramatic names we give them: death cap, deadly webcap, destroying angel, funeral bell, fool's funnel and angel's wing.

Folklore offers several ways of telling the good fungi from the bad, but none is reliable. The idea that the dangerous ones are brightly coloured probably comes from the fly agaric, which is red with white spots, everybody's idea of what a toadstool should be. It has narcotic and hallucinogenic properties and is dangerous but non-deadly. The theory that insects avoid toadstools but not mushrooms doesn't stand up to inspection. The notion that poisonous fungi invariably taste bad is also wrong: people who have survived a meal of death caps report that they were delicious; it's the after-effects that were the problem.

The death cap is, as the name implies, pretty deadly. Part of its danger comes from confusion with edible species. It can be quite large, up to 6 inches (15cm) in diameter, with colours ranging from olive to bronze. The gills (the flaps underneath the cap) are white, turning with age to cream and pink. The stems have a floppy ring around them. Their somewhat phallic look is reflected in its scientific name, *Amanita phalloides*. They contain amatoxin, which causes acute liver and kidney failure. An initial period of great pain, vomiting and diarrhoea is often followed by a false recovery, but the damage done is already irreparable and death usually follows. Half a cap can be enough.

И. БИЛИБИНЪ 1900.

There are many ways of dying, but death by fungus piques us as a matter of peculiar horror. In 2020, a case in New Zealand was reported by the British *Daily Mail*, not a paper usually noted for its coverage of New Zealand affairs. The paper told the story of a doctor and cancer survivor Anna Whitehead who nearly died after frying up a couple of death caps with fish for her lunch. She said the experience was worse than anything she had known before.

In some traditions the Buddha's last meal was poisonous mushrooms. The Emperor Claudius died, it is said, after a dish of what he thought were Caesar's mushrooms, *Amanita caesarea*, 'an edible mushroom of which Claudius was extravagantly fond', according to Robert Graves, author of *I, Claudius*. These are harmless and apparently delicious but readily confused with A. *phalloides*, the death cap. According to tradition, the deadly mushrooms were introduced to the meal by his wife Agrippina, so that Nero, her son by an earlier marriage, could become emperor in his place. Since emperors are automatically considered divine, Nero quipped that death cap fungi were 'the food of the gods'.

The Holy Roman Emperor Charles VI died in 1740 after eating death caps; the eight-year Austrian war of succession followed. The composer Johann Schobert, a harpsichordist and significant influence on Mozart, died in 1767 after a meal of death cap. He had been told they were poisonous, but, with the advice of a doctor friend, he went ahead and ate them anyway. The mushroom soup killed him, his wife, all but one of their children, the maidservant and the doctor.

The fly agaric's spectacular looks have made it the default toadstool. Psychedelic mushrooms are covered in Chapter 31, but let us add in passing that the fly agaric's hallucinatory properties are, some claim, the reason Father Christmas dresses in red and white. The sinister and shamanic association of the mushroom have been turned into prettiness in later fairy tales. The Brownies, the version of Lord Baden-Powell's Boy Scout movement designed for young girls, traditionally gather round a giant toadstool painted red with white spots. The group was founded in 1914 from popular request and was initially called the Rosebuds; this was changed a year later because the members required something a little more robust.

Perhaps the unfairest toadstools are those of the *Coprinus* genus. These are harmless enough when ingested – unless you have consumed alcohol within the last few days. The toxins attack the body's ability to process alcohol and produce symptoms like a terrible hangover.

Our horror of death by toadstool is not reflected in the number of deaths actually caused; it's much the same with deaths by sharks. The issue in both cases is the human fear of the natural world. There is still considerable danger from poisonous

Opposite *Something wicked this way comes: Russian folklore image showing evil among the toadstools; dated 1900, by Ivan Yakolevich Bilibin (1876-1942).*

fungi, but it comes on a much smaller scale: from mycotoxins found in fungal moulds. These have been met before in Chapter 31 with ergot poisoning; other mycotoxins can cause acute poisoning incidents as well as cancer and long-term damage to the immune system. The toxins can reach livestock as well as humans, and can affect meat and milk. The moulds can occur in stored food, especially in warm and humid conditions. They can damage cereals, nuts, spices, dried fruit, apples and forage. Several hundred harmful species have been identified. The World Health Organization 'encourages national authorities to monitor and ensure that levels of mycotoxins in foodstuff on their market are as low as possible'.

H. G. Wells wrote a short story, *The Purple Pileus*, in which Mr Coombes, 'a small shopman' in despair at his wife's unhelpful ways, consumes the fictional fungus of the title. Maddened by its toxins, he returns home and raises hell, but his wild rage sets all to right. At the story's conclusion, his brother makes a disparaging remark about the fungus. Mr Coombes says: 'I dessay they're sent for some wise purpose.'

SIXTY-TWO

TEAK

Wood is universally beautiful to a man. It is the most humanly beautiful of all materials.

Frank Lloyd Wright

The marvellous nature of nature has led us to believe that it's all for us: that the world is a no-strings gift to humanity. Would God have given us such fine things if he didn't want us to use them? It's a flawed argument: would God have provided psychedelic mushrooms if he didn't want us to take trips? Would he have made death cap mushrooms if he didn't want us to poison emperors? Would he have created rainforests if he didn't want us to build things from wood?

Teak, more than any other tree, can seem like something especially created for the benefit of humans. The wood from the teak tree – we are so accustomed to thinking of teak as a commodity that it seems mildly impertinent to think of it as a growing plant – is full of virtues. It's rich in natural oils which make it resistant to termites and other devourers of wood. It's hard and close-grained, which means that it endures and is weather-resistant. It has great tensile strength (resistance to breaking under pressure). If you want to build something out of wood that will last, then teak is likely to be your first choice.

The plant grows naturally in South and Southeast Asia: a deciduous tree that can reach 130 feet (40 metres). It needs good soil, warmth and consistent sunlight, which restricts it to the tropics, but it's fairly tolerant about moisture, living in areas with rainfall between 500mm and 5,000mm a year. It can cope with relatively arid areas as well as the moist forest it's normally associated with. Its yellowish heartwood darkens with age; the sapwood separates easily from the heartwood – another virtue, so far as humans are concerned. There are three species, but only one of them concerns us here: *Tectona grandis*, the tree from which all commercial teak timber comes. It was named by Linnaeus the Younger (Linnaeus's son) in his 1782 *Supplementum Plantarum*.

Teak's excellent qualities have made it a traditional building material in most of the places where it grows naturally. In India it's been used for doors, window frames, columns and beams. Entire houses have been made from teak. It's been used in boat building for at least 2,000 years, probably for a lot longer. The wood resists rot, fungi, mildew and has a very low shrinkage rate. It can be used for both framing and planking. It is very hard, but yet comparatively easy to work.

When the colonists reached Asia they took to teak in a big way. They naturally exploited the teak forests as they grew, and they established plantations, not only in the traditional areas where teak grows, but also in Central Africa and the Caribbean, and later in Central and South America. The great gift of teak timber was no longer a local matter but a boon to the whole world: and that's how it remains: a wood that comes from both managed plantations and forests, grown all over the tropics and exported all over the world.

It's widely accepted that the destruction of rainforest is a bad thing. But to most people, at least in the market countries, this is something of an abstract idea: a pure virtue pursued by the self-righteous. There is a lack of fastidiousness about the provenance of tropical hardwoods. WWF surveyed this issue in the UK and reported: 'With some notable exceptions, most retailers don't appear to know or care where the wood in their products comes from or if it is responsibly and sustainably sourced.' This can only be because their customers aren't interested either: things would soon change if customers routinely walked away from doubtful wood.

Teak products are widely on sale: often used for indoor and outdoor furniture as well as for boat building. The ethical issues around its use come down to the individual who believes that one silly old bench won't make much different to the future of the earth. In one sense there are no clear answers to this. Sustainable timber comes from plantations. These may well be established on land that was once virgin rainforest, but at least timber from such sources is not making things even worse. But a good deal of teak still comes from wild forest. This is extracted under licences granted to logging companies: and at once we are talking about a system that is open to corruption. Even when such licences are above board, the practice of destroying forests can continue illegally: logging generally takes place in remote places, with only the local people to suffer.

The idea of illegal logging sounds engagingly piratical: intrepid bands of rascals making an outlaw living by their boldness. But destroying – and, more particularly, transporting – enormous trees is an enormous undertaking: it only works if it's heavily capitalised: in other words, it's a corporate crime. This contribution to the ongoing ecological holocaust involves massive violations of human rights, both in the labour and in the indigenous people whose forests are being destroyed. In Myanmar teak has been cut and sold in immense quantities after international aid was cut off following the slaughter of pro-democracy demonstrators in 1988.

This has given rise to the expression Conflict Teak: wood that reaches the developed world by way of environmental destruction and political corruption. I should point out that not all tropical hardwood products involve this combination of horrors: just that people in the market countries are insufficiently aware of the

Opposite Best wood: teak loaded at Kokogon, Burma, a 1905 painting by Robert Talbot Kelly (1861-1934).

difference between Conflict Teak and the other kind. Major financial institutions have been accused of involvement with this profitable trade: mostly by not looking too closely. The World Bank has estimated the annual value of the illegal trade in timber – which involves many species – at US$10 billion; this brings in an additional loss to the governments concerned of US$5 billion. Forget about rascally land-pirates: this is a major international illegal business, to be compared with the trade in drugs or arms.

And it happens for the same reason: the product is vastly desirable. Teak is the finest wood around: for many purposes it's the best possible material. Where there are people willing to buy and people willing to sell, there is always a market. The answer, insofar as there is an answer, lies in the customer. No customers, no market.

SIXTY-THREE
PINE

They kill good trees to make bad newspapers.

Environmentalist slogan

I have worked as a writer for half a century: how many trees have I killed? All those books: the more they have sold the happier I have been and the more trees have been cut down. And all those newspapers and magazines I have contributed to, day after day, year after year: my words recorded on stuff made from trees, to be read or not read and then discarded. How many of these trees came from sustainable and non-damaging forestry?

Most of these will have been pine trees: the family *Pinaceae*, which includes the genus *Pinus*, also spruces in the genus *Picea*: plants much grown in commercial plantations. On my shelves are the works of many great conservationists, many of whom would have reservations about commercial forestry.

Humans have used paper for more than 2,000 years; wasps a great deal longer. Wasps use chewed wood pulp to make their exquisite paper nests. The human given the credit for the invention is the Chinese court eunuch Cai Lun, who came up with it in 105 BC. Paper can be made from any kind of wood, also from hemp, flax and cotton; it can also be made from waste paper and rags. These days it is mostly made from trees from the pine family.

The trees are debarked, chipped and then pulped. This process removes the cellulose fibres from the lignin; lignin is the substance that allows trees to become rigid, but it's not good for paper. After that, sheets are formed as the fibres are flattened and water is removed. What's left is dried further at temperatures above 100°C (212°F). In 2018 the global industry produced more than 400 million tonnes of paper and cardboard, of which more than half was used in packaging.

Pine trees are evergreen and cone-bearing. The word pine is used informally for many species outside the *Pinus* genus. They are all gymnosperms, the second great plant group after flowering plants or angiosperms. The *Pinus* genus includes some of the oldest trees on the planet: the Great Basin bristlecone pines; one example called Methuselah in the White Mountains of California is 4,600 years old. Pines carry both male and female cones; the male cones are small and last only a short time, spreading their pollen which is distributed by the wind. The female cones take on the pollen but delay their fertilisation until the following spring; it can

Soft wood: Forest, 1898, by Ivan Shishin (1832-98).

take a cone up to three years to mature after pollination. In most species the resulting seeds are winged and dispersed by the wind. The gymnosperms are about twice as old as the flowering plants; pines have been growing on earth for at least 200 million years. They are mostly native to the northern hemisphere, but a few extend below the equator. They are capable of living in the coolest and the hottest places where trees can grow at all, and are found from sea level up to 17,100 feet (5,200 metres).

But this capacity for extremes is not what has made them so attractive to us. It is the fact that temperate species of pine are fast-growing softwood trees which are useful to humans in many different ways, including the manufacture of paper. The timber is used for furniture, window frames, floors and roofs. It is easy to work with and yet pretty durable. It makes good-quality sawing timber, but we no longer work only with the wood as it comes from the tree. Plywood uses veneers of wood in layers with their grains running at 90 degrees to each other for strength; softwood plywood, made from pine, spruce and some other woods, is important in construction. Various forms of composition board are made from cellulose fibres and woodchips held together with adhesive. The original tree no longer dictates the form of the timber.

Timber is hugely useful stuff, but when you make too much use of it you run out. Britain, along with many other countries, realised that after the First World War and started to establish plantations of pines and other conifers on soil that was too poor for agriculture, in places like the New Forest and Thetford Forest, and then into the uplands, with Kielder Forest. These plantations, often filled with non-native trees, all the same species, all the same age, existed purely for the crop. This was a continuing trend for many years: in the 1980s the internationally important peatlands of the Flow Country were ploughed up and planted with exotic pines and sitka spruce, a project backed by government grants and tax breaks.

This intense farming of trees makes them vulnerable to the usual dangers of monoculture. The plantations of lodgepole pines in Canada might have been set up to please pine bark beetles. The relationship between tree and beetle has continued across the centuries: the beetles get beneath the bark, which is there for the tree's protection, and into the softer layer beneath to lay their eggs. When they do so they bring in a few spores of blue stain fungus. This fungus develops and starts to feed on the wood of the trees; by the time the egg has hatched the tree's tissues are available to the larva as a nutritious gloop, brought into existence by the digestive processes of the fungus. The tree is attacked by a fungus that eats it, and at the same time, the beetles damage its vascular system. It's a tough challenge, but over the centuries the trees have coped and the balance has been maintained. But things changed when plantations were established. There are now many more beetles, because there are many more lodgepole pines. With climate change many more beetles survive the winter; climate change is particularly severe on monocultures, because monocultures have no resilience. The lodgepole pine is a classic example of a planet with no Plan B.

Travel through Scotland and much of the landscape is characterised by bare hills: many romantic songs have been written about them:

> And fair as these green foreign hills may be
> They are not the hills of home...

But these hills were once forests, dominated by Scots pines. They were felled for the usual reasons – timber is too useful to leave on trees – and the hills have been kept bare by sheep and huge herds of deer; the deer have no natural predators since wolves became extinct in Scotland in the seventeenth century. They are maintained in large numbers so that rich people can shoot the mature stags. The deer eat any shoots of pine that raise their heads above the low sward.

But in some places the Caledonian Forest is being regenerated by keeping the deer out and allowing the trees to regenerate. I have been to Abernethy in the Scottish Highlands and seen the new growth of pines rising like a spiky lawn; I have visited Alladale, where getting on for 1 million trees have been planted in the once-naked hills. The owner, Paul Lister, dreams of one day repopulating the land with wolves.

These days commercial planting and government policy on forestry is no longer bleakly utilitarian. It's accepted that plantation forests should play an important role in the maintenance of biodiversity and the regulation of climate. There is a growing concept of what is now called 'natural capital', and it's clear that the timber industry needs some kind of environmental ethic for its long-term survival. It's not just the industry that benefits from an ethical approach: environmental care is ultimately about the survival of all species on earth, including our own.

SIXTY-FOUR
MULBERRY

We are all Adam's children but silk makes the difference.

Thomas Fuller

Mulberry was used in the making of the earliest paper, after its invention in China in 105 BC (see previous chapter). The world's thinnest paper, *tengujo*, is made in Japan from mulberry; the length of its fibres gives it extraordinary strength despite its thinness. It is used for the conservation and restoration of documents, books, paintings and statues.

Mulberries are tall deciduous trees that grow wild in temperate climates over most of the world. Their classification is complex and disputed; some put more than 200 species in the *Morus* genus. Three of them have been widely cultivated for their berries and are named for the colour of the fruit found on their best-known cultivars: white, red and black. The wild versions of the white mulberry actually produce black fruit. A single tree produces a bonanza: blackberry-like fruit in thousands. Mulberries have been banned in some American cities, partly because the fruit make a mess, both as they fall when they turn up in bird dropping; and also because they produce a great deal of fine pollen, particularly disagreeable for people with pollen allergies.

They have been cultivated for their fruit across the centuries; they were banquet-quality food for Ancient Greeks and Romans. They were also prized for their healing qualities; the Romans used them for diseases of the mouth, throat and lungs, Native Americans for dysentery. They have also been used to treat ringworm. Mulberry trees have often been planted in infirmaries and sanatoriums. The white mulberry fruits are praised for their tart flavour, black for their greater depth. White mulberries are native to Southeast Asia, but humans have spread them more or less anywhere they will grow. They have been taken into the Americas; in the United States the white mulberry now dominates the native red. A cold-tolerant Russian cultivar was introduced to the United States, more for shelter and timber than for its fruits; it's now regarded as an invasive species. Both the fruit and the leaves are considered a useful dietary supplement for health-conscious people. Juice from the fruit makes a natural food colouring. A cultivated variety of weeping mulberry is a popular 'lawn tree' in parts of the United States.

Opposite All for silk: Picking Mulberry Leaves, *Japanese woodblock c.1800 by Kitagawa Utamaro (d.1806).*

The mulberry tree supplies the pay-off in the story of Pyramus and Thisbe, which is best known in the English-speaking world as the subject of the play performed by Bottom and the rude mechanicals in *A Midsummer Night's Dream*. The clowns don't mention the mulberry, but Ovid gives it full value in *Metamorphoses*. The couple are Babylonian lovers who do their courting either side of a wall. They agree to meet at night beneath a mulberry tree. Thisbe arrives first but is frightened off by a lion. Pyramus arrives a little later and, in the belief that the lion has killed and devoured his beloved, takes up his sword, and kills himself. Thisbe returns from hiding, finds the body and takes up the sword to turn on herself:

And you, O tree...
Preserve the marks of our death; let your fruit forever be dark
As a token of mourning, a monument marking the blood of two lovers.

The gods' hearts are touched and the mulberry is now the colour of blood.
The mulberry also appears in the children's rhyme:

Here we go round the mulberry bush
The mulberry bush, the mulberry bush...

It is a learning song: 'this is the way we wash our clothes' and so forth. Vincent van Gogh painted an autumnal mulberry tree that grew in the grounds of the asylum where he lived in 1889, as powerful a picture as he ever painted.

There is one other significant fact about the mulberry: it is the sole food-plant of the caterpillars of the moth species *Bombyx mori*; the second (specific) name referring to the genus *Morus*. The white mulberry *Morus alba* is their preference, but they can subsist on related species.

This reliance on a single species or a very small range of species might seem like an evolutionary mistake: an impossibly finicky way of living. But there are important advantages in specialisation. First, tying your life to a single plant allows you to coordinate the timing of egg-laying and hatching with the appearance and growth of the host plant, and so to establish an advantage in places where that plant thrives. The second concerns the perpetual evolutionary arms race between plants and their predators. A generalist, a species that eats many kinds of plants, has a good deal of choice – but an equally considerable number of species will be unavailable, because of the defences they have evolved. A specialist keeps pace with the adaptations of the plant. The generalist takes many species – but that leaves an opportunity for others to exploit the plants that the generalists can't cope with.

Butterflies and moths divide their lives very precisely: adult life is for reproduction: larval life is for eating. But it's the in-between stage of *Bombyx mori* that interests humans: before the caterpillar becomes a pupa, it spins itself a cocoon of silken thread. This is the silkworm: the creature that created a luxury industry that has thrived for 5,000 years and built one of the most important trading routes in history. The caterpillars can take most of the credit for this, but they couldn't have done it without mulberries.

The story of the discovery of silk is of an accident: a cocoon that fell into a cup of tea being drunk by the mythological Chinese Empress Leizu. Once dunked, it unravelled. The fabric created from the silk was good to wear and good to look at. Still more important, it indicated social status. Silk remained a Chinese monopoly until AD 500, and was a fiercely protected manufacturing process. The Silk Road opened up around the second century AD and linked cultures and communities along its route from China into Europe. It lasted until the Middle Ages; when sea trade, though still dangerous, was becoming more viable.

Silk had a European reboot when the returning crusaders brought the material home. It was obvious that if you could make silk in Europe, you would be on to a good thing. The problem is that you can't make silk without mulberries. Silk moths are wholly domesticated creatures. The caterpillars have no defence mechanisms and can't survive in the wild; the adults can't even fly. They have been selectively bred to produce far more silk per cocoon than the original wild species. The silk moth has been bred for human convenience: but the caterpillars still can't survive without mulberries.

The plants can be hard to grow in cold places: but they were cultivated in Calabria, right down in the toe of Italy, and so European silk manufacture began in Italy. Cultivation of more hardy mulberries that were still palatable to the caterpillars was important; more tolerant varieties were developed for more northerly places. James I encouraged the planting of black mulberries to establish a silk industry in Britain: he planted 4 acres at what is now Buckingham Palace. Mulberries can still be found in royal parks, including St James's Park and Windsor Great Park. Silk has always been associated with royalty and with just about every other kind of elitism: silken gowns and silken shirts for the great public occasions, silken sheets for the great private ones. Silk is associated with artifice, glamour, expense and the Great Indoors: and as you inspect the swatches of silken garments on sale beneath the artificial lights of every airport lounge, it can be hard to accept that the entire industry is dependent on a certain kind of tree.

SIXTY-FIVE

LILY

'And do you know what your job stinks of, you dumb bastard?' he
asked cheerfully. 'Formaldehyde and lilies.'

Felix Leiter to James Bond in Ian Fleming's *Diamonds
Are Forever*

Flowers are about life, so it is inevitable that they are also about death. Across the millennia we have used flowers to honour the dead, to express sorrow and love, to stand for the hope and belief in life's continuation and to mask the odour of decay. In much of Europe and North America, the flowers most often chosen for the job are lilies.

Many flowers are called lilies: waterlily (see the next chapter), lily of the valley, day lily. In this chapter I am mostly referring to plants in the genus *Lilium*, and in particular to the Madonna lily, *Lilium candidum*. There are about 100 species in the *Lilium* genus; they grow wild in the temperate regions of the northern hemisphere, reaching between 2 and 6 feet (60-180cm), and are found mostly in woodland and mountains. They are perennials, rising each year from bulbs, so they are no trouble to grow: just bury them in the dormant season and wait for the following year, when they will put on a show of extravagant, colourful, aromatic flowers. They do well in pots, too: obliging plants that reward minimal care with lavish blooms.

There are many hybrids with spectacular forms and colours. The bulbs are eaten as a root vegetable in Chinese and Japanese cooking. The powerful scent and the size and colour of the flowers evolved in wild species to entice pollinating insects. What pleases the bees also pleases humans. They are perhaps one of the oldest plants to be cultivated for their flowers, and selective breeding has brought about exaggerated size, shape, colour and aroma.

The mutual dependence of many flowering plants and their pollinators rightly fascinates us. But it's also worth noting the mutual dependence between flowering plants and humans. We have met this dependence already in these pages, with the food-plants we cultivate: the plants feed us and so we labour for the plants' benefit. But there's a similar relationship with plants that are merely (merely?) decorative.

Opposite Prince of Lilies: *relief fresco c.1550 BC from the palace at Knossos, Crete.*

Lilies for sanctity: The Annunciation, c.1485-92, *by Sandro Botticelli (d.1510).*

Many species of plants, lilies and, most obviously, roses have benefited because they give not nutrition but sensory delight to humans. As a result we make space for them, remove their competitors and attend to their needs. This is not a one-way process: we look after such plants and, in return, they give us a beauty that seems necessary for our lives, perhaps now more than ever.

Flowers are about love and beauty, not small things in any life. The Victorian language of flowers assigned different meanings to different flowers so that they could be sent as a gift with an implied message: one that could not, perhaps, be spoken out loud. A gift of lilies was an expression of love and ardour that could be made with perfect propriety; orange lilies gave a message of happiness and warmth. Lilies are traditional funeral flowers: a plant that can give a little comfort in times of grief has immense value.

Humans have been putting flowers on graves for perhaps 70,000 years. The Shanidar Cave in Iraq contains ten human skeletons; one at least has been adorned with flowers. Pollen from twenty-eight species has been found in the cave; the concentrations are higher in the graves than elsewhere in the surrounding area. The Raqefet Cave in Israel has four graves 12,000 years old, unambiguously adorned with flowers. It's possible that the Natufian people who performed these burials were the first humans to abandon the nomadic life and settle down in permanent dwellings; a move that would require a permanent graveyard. The tradition of adorning such places with flowers has continued more or less unbroken. Egyptian pharaohs were buried with flowers; the mummy of Rameses II was decorated with thirteen rows of floral garlands. When there is death, we need flowers.

The Madonna lily has long been the first choice for this in Europe and North America. This species, much cultivated, is also an impressive wildflower, native to the Balkans and West Asia. It is unmistakably depicted in a fresco at Knossos in Crete called *The Prince of the Lilies*, a famous and enigmatic image of a powerful male figure against a background of lilies, dating back to *c.*1550 BC. There are lilies in the biblical text the Song of Solomon (see also Chapter 3): 'As the lily among thorns, so is my love among the daughters.' The *lilium inter spinas*, the lily between thorns, has become proverbial, though there is dispute as to whether the original Hebrew word *shoshannah* should be translated as lily or rose. Either way the lily has been revered as a flower packed with sacred meaning, particularly associated with the Virgin Mary since at least the Middle Ages. In many representations of the Annunciation the archangel Gabriel, who tells her that she is to bear the son of God, is handing her a Madonna lily, perhaps most vividly in Botticelli's version. The flowers are white and therefore associated with chastity; their beauty is clear to all, as is their fragrance. They have also been called white, French, annunciation, St Joseph's and Bourbon lily.

In a few churches in England you can find depictions of Christ crucified on a lily. One possible reason for this strange image is that the annunciation and the crucifixion are said by some to have taken place on the same date, 25 March (nine months before Christmas, the ceremony of Christ's birth). Easter is particularly associated with lilies: it is a festival of spring and therefore marks the rebirth of the year as well as that of the risen Christ. The church is decorated with flowers, which in earlier centuries were mostly unavailable in the previous months, and lilies were the most favoured. This association with the festival of rebirth makes them especially appropriate for funerals.

Lilies also have some less pacific meanings. A stylised Madonna lily, the fleur-de-lys, was the badge of the French royalty, though some say the flower on the shield is more like a flag iris. A religious symbol was appropriated for secular purposes, implying that its bearers had a God-given status: a divine right to rule. If the Virgin Mary was backing you, what could go wrong? There was also a lily on the banner of Saladin, who defended what was called the Holy Land against the invading crusaders from Europe.

The lily remains one of the most popular cut flowers; you can buy them at supermarkets. Not everyone is happy to give or receive them as gifts, because of their funereal associations. Nor does everyone like the heady scent of lilies in small enclosed spaces. But the scent is a reminder of at least one of the reasons flowers in general and lilies in particular are associated with funerals. They do a better job than practically anything else in the natural world of masking competing smells. A fresh corpse has no chance against a garland of lilies.

SIXTY-SIX

WATERLILY

'"Water lilies" is an extension of my life.
Without the water the lilies cannot live,
as I am without art.'

Claude Monet

The River Waveney divides Suffolk and Norfolk in the east of England. There's a place where the river slows almost to a stop and in season you can canoe through waterlilies, green leaves and yellow flowers, an expanse of almost ridiculous beauty. Here is a group of plants that inspired one of the greatest painters, gave the idea for one of history's most significant buildings and might be considered the solution to what Charles Darwin called 'an abominable mystery'.

There are about sixty species in the family *Nymphaeaceae*; the lotus (Chapter 38) is in a different family. They grow wild across tropical and temperate climates and most are perennials. They all have waxy leaves full of air-spaces, so they float well. They grow on long stalks from rhizomes buried in the mud and produce single flowers above the water: flowers so wonderful that you fear they will catch Richard Dawkins off guard and turn him into a true believer.

Every botanical garden must have its lily-pond. They're not hard to cultivate: many private gardens have a small pond with lilies; I did so myself until benign neglect turned it into a bog-garden. Grand gardens in fine houses seldom lack a pond full of lilies: they are plants that give unambiguous delight.

They look like the most subtle and sophisticated plants, but they are an ancient lineage – though the question of ancientness is a vexed one. About 90 per cent of all plant species on earth today are angiosperms: flowering plants. But the fossil evidence for their existence doesn't go as far back as you would expect: in Charles Darwin's time the oldest fossilised flowering plant came from the Cretaceous period, towards the end of the time of the dinosaurs; it is about 100 million years old. Before that, the earth's terrestrial flora were dominated first by clubmosses and horsetails (see Chapter 16) and then by gymnosperms, mostly conifers. Flowering plants appeared with apparent suddenness alongside the dinosaurs, and seemed to take over in double-quick time.

__Opposite__ Floating wonders: Waterlilies, 1916-19, by Claude Monet (1840-1926).

This didn't fit Darwin's theory. 'As natural selection acts solely by accumulating slight, successive, favourable variations, it can produce no great or sudden modifications, it can only act by very short and slow steps,' he wrote in *On the Origin of Species*. But twenty years later he famously wrote to his lifelong friend, the botanist Joseph Hooker, that the appearance and spread of flowering plants was 'an abominable mystery'. His best suggestion was that earlier fossils would someday be found in some unexplored region.

Since then, earlier fossils have been found in China, predating the earliest fossils Darwin knew by up to 35 million years, so Darwin was at least part right. Fossils of similar age have also been found in Europe and North America. The fossil record is always patchy and capricious: even if we found all the fossils in the world, they wouldn't tell us a complete story because it's extremely hard for any living thing to become a fossil, especially if it's soft. But genetic analysis of waterlilies has hinted at great age: they are what students of classification call basal angiosperms.

Other genetic studies, notably on the strange *Amborella* of New Caledonia, look back towards the beginning of flowering plants, though this one can't claim to be the ancestor of all flowering plants: it's more of an outlier, in the manner of the platypus. Statistical analysis of fossils has given a speculative date for the first angiosperm at 250 million years ago. The waterlily plays an important part in the study of the evolution of angiosperms: yet another living thing telling us that Darwin was right all along. The abominable mystery of the rise of the angiosperms is not yet settled for all time, but with the help of the waterlily and others, the scientists are getting there.

The largest waterlily species are in the genus *Victoria*, chosen to celebrate the woman at the head of the British Empire. There are two species, the Amazon or royal waterlily, and the Santa Cruz. There was naturally a nineteenth-century status race to grow them in England, and the dukes of Northumberland and Devonshire were in the lead. William Cavendish, the 6th Duke of Devonshire, had the nous to employ Joseph Paxton, whom he poached from the Royal Horticultural Society. Paxton succeeded in growing the plant, no easy matter with coal-fired heating, and the resulting bloom was duly presented to the lady it was named for. Paxton was an architect as well as a gardener, and built a larger glasshouse to accommodate the lilies, which was modelled on the structure of the leaf of the lily.

This is a remarkable thing, effortlessly 6 feet (nearly 2 metres) across, sometimes larger. Photographs have been taken of the leaves with a child sitting safely on them: there is one showing Paxton's daughter Alice doing just that. The flower is a monster, up to 18 inches (46cm) across; Her Majesty must have been delighted with her gift. The leaves maintain their strength, despite their comparative lightness, by means of a ribbed under-surface: radiating ribs linked by flexible cross-ribs, which Paxton termed 'a natural feat of engineering'.

He developed this system in his design for the Crystal Palace: the vast steel and glass building that housed the Great Exhibition in Hyde Park, London in 1851, taking advantage of advances in the technologies of glass and cast iron. The building covered 990,000 square feet (92,000 square metres) – or about four times the size of St Peter's in Rome. The exhibition and the building that housed it were perhaps the peak, the perihelion, the greatest moment of glory of the British Empire. Paxton also developed a variety of banana and named it for his employer's family: the banana that dominates the world today, the Cavendish (see Chapter 58).

Claude Monet, the great impressionist painter, said: 'My first masterpiece is my garden.' It is still there at Giverny in Normandy, north of Paris. A river was diverted to create the watery landscape he wanted, for the reflection of light on water had him in thrall throughout his life. He planted bamboos, weeping willows – and waterlilies. And then he painted them: there are at least 250 paintings of waterlilies in this sequence. He had already shown a taste for returning to the same subjects again and again in different lights and moods: in later life he took this to its logical conclusion: 'This is beyond the strength of an old man and yet I want to express what I feel,' he said.

These paintings increasingly became works of pure colour, as he developed a cataract condition in his eighties. He died a few months before a display of his waterlilies went on show at what is now the Musée de l'Orangerie in Paris. They were presented by Monet to the state of France to mark the end of the First World War: eight vast canvases with a total length of almost 330 feet (100 metres): as much an installation, as such things would be termed a century later, as pictures in frames. They are a double celebration: peace and beauty by way of waterlilies.

SIXTY-SEVEN

ORANGE

He hangs in shades the orange bright,
Like golden lamps in a green night…

Andrew Marvell

The Louvre Palace in Paris was the principal dwelling of the kings of France, so naturally it had an orangery. In 1682 Louis XIV moved the royal household to Versailles, leaving the old palace for the royal collection; it's now the Louvre museum. As we saw in the previous chapter, the orangery now houses eight of the great waterlily paintings by Claude Monet. Versailles also has an orangery, of course, and it's a belter. The passion for oranges inspired the construction of buildings and annexes with as much glass as possible, so that orange trees could have the warmth and light they need to thrive. This idea led eventually to the glory of London's Crystal Palace (also in the previous chapter).

Orange is not a species. The familiar sweet orange fruit grows on a tree that is a hybrid of pomelo and mandarin, which are both plants in the *Citrus* genus. The tree produces a fragrant flower and a modified berry, technically a hesperidium, which is divided into segments or carpels. In the 1980s oranges became the most cultivated fruit trees in the world.

They are thought to be native to the Malay Archipelago; they were cultivated early in the southeast Himalayan foothills. The sweet orange is the variety that has always dominated, though there are plenty of others, including mandarin, bitter or Seville orange and bergamot. Other cultivated trees from the *Citrus* genus include lemon, lime and grapefruit. There is a reference to the sweet orange in a Chinese document dated to 314 BC. They spread westward with the Arab Empire, which went as far east as the Chinese border. Oranges were introduced to North Africa, Sicily and the Iberian Peninsula. Large-scale cultivation in Europe can be dated to the tenth century. A document of 1475 written for Pope Sixtus differentiates between sweet and bitter oranges. By the sixteenth century they could be found all around the Mediterranean.

They require a fair amount of warmth to do well, and though some say that a light frost is good for the tree, they can't take much of it. That made them a luxury item in North Europe. The explorers of the Age of Discovery took them to the Americas: Columbus planted them on Hispaniola, now Haiti, on his second

Gracious living: The Orangery in the Ducal Park, *1857, by Guido Carmignani (1838-1909).*

voyage of 1493. Franciscan monks took them to California. Dutch, Portuguese and Spanish sailors planted them along the trade routes they established, because by then it was known that oranges could prevent and cure scurvy. It's been estimated that in 300 years of sea travel from 1500 to 1800, 2 million sailors died from scurvy. The Spanish were aware of a cure in the sixteenth century: in 1579 Agustín Farfán, a Spanish friar and physician, recommended oranges and lemons. In 1747 James Lind, a Scottish doctor and pioneer of naval hygiene, performed one of the first clinical experiments in the history of medicine and demonstrated beyond doubt that orange juice – he recommended that it be mixed with white wine or beer – was effective against scurvy. Ocean travel grew safer because of oranges.

The taste for oranges drove the creation of orangeries. The first orangeries were built in Italy, where glass technology had reached the stage where large expanses of clear glass could be produced. They were initially strictly practical buildings for producing fruit, and were heated with open fires; one was built in Padua in 1545. It was a couple of centuries before the technologies permitted a glass roof, but the charm of an indoor grove of trees was irresistible and, perhaps more importantly, an orangery became a status symbol.

If you have a status symbol you need to show it off to people, so an orangery became a place not just for arboriculture but for visits and entertainments, often furnished with grottoes and fountains. Louis XIV, with his taste for both ostentation and unusual food, naturally had to have the best: the result, at Versailles, is 508 feet (155 metres) long and 42 feet (13 metres) wide. I remember walking it as a teenager, a little queasy that you could see for such a distance while still indoors. It contained 3,000 orange trees in pots of solid silver: a fact that helps to explain the Revolution that followed in 1789.

One reason an orangery was such a glorious extravagance was window tax. It was considered that a tax on income would be bad manners by the state, since a person would have to reveal the extent of the income enjoyed. The answer was to tax the dwelling: and if you counted the number of windows you could get a reasonably honest indication of wealth. Such a tax was imposed in both England and France: in buildings of the late eighteenth and early nineteenth century you can often see houses of elegant shape with bricked-in windows, an early form of tax avoidance. It followed that an orangery was an insolent display of wealth. (More on the tradition of the hothouse and the greenhouse in Chapter 69 on cucumbers.)

It was a tradition that began with oranges: the chance to enjoy pleasant and exotic fruit and to show off: an irresistible double. But with increasingly efficient transport oranges gradually became democratised. They were one of the treats greatly missed by the British public during the Second World War, when most international trade was impossible.

Orange production was drastically affected by an insect called the cottony cushion scale, which reached American orange groves from Australia in 1868. This wiped out the orchards around Los Angeles and affected all California. It was eventually controlled by introducing an Australian ladybird species that predates the invading insects.

Trade in oranges is complicated by the fact that they won't ripen if you pick them early. They must travel fully ripe, but they keep better than most fresh fruit. They often have patches of green and yellow: these can be removed by treating them with ethylene gas in a process called degreening. The trees are largely infertile, mostly propagated by grafting. They are tolerant of many different soil types and can be cultivated almost anywhere the climate is right.

The demand for oranges increased dramatically in the 1920s when people started to drink the juice. This is high in vitamin C, but also in sugars; it's not the unambiguous health-giver it was once thought to be. A good deal of what is sold is made from concentrate: if you take most of the water out, it is much easier to transport. As a result, juices that are 'not from concentrate' have a premium value. In 2017, 73 million tonnes of oranges were produced across the world; Brazil grows nearly a quarter of the world crop, followed by India and China.

Before people were familiar with oranges there was no word for the eponymous colour. It was called red-yellow, or, more exotically, saffron (see next chapter). Did the lack of a word affect the way we perceived the world? But by the time Newton listed the colours of the rainbow (see Chapter 45), he had the fruit for reference.

SIXTY-EIGHT

SAFFRON CROCUS

In saffron-coloured mantle, from the tides of oceans rose the
morning to bring light to gods and men.

Homer

Like orange, saffron is a colour as well as a plant, or part of a plant. Saffron has long been associated with luxury: by weight it's the world's most expensive spice, not least because its preparation is highly labour-intensive. But it's also associated with religion: more, with holiness, purity and the most sacred things of all.

There are 249 accepted species of crocus, but only one produces saffron: *Crocus sativus*. The substance traded as saffron comes from the pistil, the female parts of the flower. These comprise the stigma, which receives the pollen, and the styles, which connect the stigma with the ovary. They give out a fragrance usually described as hay-like. It's produced by phytochemicals: notably picrocrocin and safranal; they contain in total twenty-eight volatile and aroma-yielding compounds, along with carotenoid pigments which give a golden yellow colour to both food and textiles.

The plant probably originated in Iran, though there are claims for Greece and Mesopotamia; the Greek word *krokos* comes from a word from the Semitic language group meaning saffron. The plant is unknown in the wild today; its nearest relative is the plant known as wild saffron, *C. cartwrightianus*. The cultivated saffron crocus is incapable of independent sexual reproduction and is propagated vegetatively, by dividing the bulbs the plants spring from.

Finicky task: a saffron gatherer, c.fifteenth-century BC *fresco from the palace at Knossos, Crete.*

The harvesting of saffron is a delicate business requiring many hands. It's been estimated that 200,000 stigmas from 70,000 crocuses make 1 pound (0.45kg) of finished product. A single crocus will produce, when dried, 7 micrograms (0.00025 ounces) of saffron. The quality depends on the ratio of stigmas to styles: the more stigmas the better. Naturally there is a long history of adulteration with fibres of beetroot and pomegranate, and strands of silk. Tasteless and odourless stamens (male parts of the flower) might be included. Colouring obtained from gardenia fruits has been used to forge saffron.

The substance has been treasured across the centuries. It's mentioned in an Assyrian botanical work of the seventh century BC. Alexander the Great used to bathe in it: it was said to be good for wounds. It could be scattered onto a bed as a cure for melancholy. Cleopatra also bathed in it: a saffron bath was said to increase the pleasures of love. The substance has been celebrated as an attribute of the dawn in Classical works; the goddess of the dawn is Eos in Greek, Aurora to the Romans. In the *Iliad* Homer writes of the saffron dawn and is echoed by Virgil in the *Aeneid*: 'Aurora now had left her saffron bed...'

Saffron has most often been used as a costly and exotic addition to food. The Ancient Romans used it; it was said to have the additional virtue of alleviating problems of digestion. It continues to be used in cooking in many different cuisines for the combination of its delicate aroma and joyful colour: paella in Spain, the fish stew bouillabaisse in the south of France, risotto Milanese in Italy, pilaf dishes in west Asia; in Indian cuisine it is mainly used in puddings, particularly those made with milk and rice. You are recommended to soak the threads in hot liquid – water, stock, wine – for a couple of hours to release the flavour. The fault is no doubt with my own palate, but I've always found saffron a little disappointing, both in food I have cooked and food that has been cooked for me. It's enough to make you wonder if the big deal about saffron is more to do with price than flavour.

Saffron is produced from Spain to Kashmir, though 90 per cent of the world's supply comes from Iran. It was produced in Britain in the sixteenth and seventeenth century: the name of the Essex town of Saffron Walden celebrates the fact that it used to grow the spice; the trade was revived there in 2013. The charisma that the spice possesses is perhaps related to the colour, and in some traditions to the way the colour is venerated. Monks and nuns, holy men and saints in both Hindu and Buddhist traditions traditionally wear the saffron robe.

This is not just because saffron smells nice. Yellow, red and orange are the colours of fire, and fire is central to both religions (we have already encountered Buddha's Fire Sermon in Chapter 30 on the Bodhi tree). Fire consumes everything and everyone, both rich and poor. We will all be destroyed: realisation of that truth is essential to self-knowledge. Fire is about sacrifice, renouncing all that you have to the consuming flames. Those who have renounced everything are entitled to wear the saffron robe. Fire is also abstinence, cleansing and purity. In the West, fire is associated with ungovernable passion; in these religions fire is about the conquest of such things.

When Mahatma Gandhi was working for Indian independence from the British rule (eventually realised in 1947) he proposed a flag for a united India, one in which groups at odds with each other would be reconciled. Accordingly, the flag is a tricolour, with green, the colour of Islam, at the bottom and saffron, for Hinduism, at the top. The white stripe in the middle stands not only for all the other smaller groups, but for peace between them all.

The association of saffron with Eastern religions made it attractive to the hippy movement of the late '60s and early '70s. It was celebrated by the Scottish singer-songwriter Donovan, who was at one stage considered Britain's answer to Bob Dylan:

I'm just mad about Saffron
Saffron's mad about me...

Saffron became a name with vaguely hippy connotations: in the British sitcom *Absolutely Fabulous*, the lapsed hippy Edina has a daughter named Saffron, called Saffy.

The price remains the most remarkable thing about saffron. Saffron robes were seldom dyed with real saffron: it is cheaper and easier to use turmeric, which can be obtained at reasonable cost and is made from tubers, the most robust rather than the most delicate part of a herbaceous plant. But the association with the more exotic spice matters: even to those who have renounced everything, it is better to wear the saffron robe than the turmeric robe.

SIXTY-NINE
CUCUMBER

A cucumber should be well sliced, and dressed with
pepper and vinegar, and then thrown out,
as good for nothing.

Dr Samuel Johnson

The great divorce between humans and nature began 12,000 years ago with the beginning of agriculture. The pace hotted up about 10,000 years later because the Roman Emperor Tiberius had a taste for cucumbers. The need to supply the Emperor – described by Pliny the Elder as 'the gloomiest of men' – with a daily ration of cucumbers was the beginning of the divorce of agriculture not just from the weather but from the seasons.

Tiberius, Rome's second emperor, who reigned AD 14-37, left Rome in AD 30 and set up on the island of Capri. Physicians told him that he needed to eat at least one vegetable daily, and he had a fancy for cucumbers. The trouble is that it's hard to grow cucumbers in winter, even on Capri. But you pull out the stops for an emperor with powers of summary exaction, and the gardeners came up with wheeled carts, that could both follow the sun and go under cover on cold days. They improved this with light-permeable coverings: the best of these was made from selenite, a stone that could be polished to translucence. They had invented the greenhouse. They had taken a giant step towards the conquest of the climate.

The cucumber is a creeping vine with origins in Southeast Asia. It produces a type of berry known as a pepo, which has a hard outer rind and no internal divisions. It is about 95 per cent water. Charlemagne was also keen on them. They came to England in the fourteenth century but didn't catch on; all the same, Christopher Columbus took them to Hispaniola, now Haiti, in 1494. They were reintroduced to England in the seventeenth century and had a mixed reception. People thought that eating uncooked vegetables would lead to disease – quite right too, if they're not properly cleaned in safe water. Some said the vegetable was fit only for cows, and so they were called cowcumbers, though the word cucumber has a respectable Latin origin. Samuel Pepys wrote in his diary: 'This day Sir W. Batten tells me that Mr Newburne is dead of eating cowcumbers.'

In northern countries you can only reliably grow cucumbers by altering their environment – and that gave them a certain cachet. The idea of raising plants under glass was spreading. The use of heated greenhouses is mentioned in a Korean text of 1450, and there are early examples of the same principle in the Netherlands and England. In 1577 Thomas Hill published *The Gardener's Labyrinth*, in which he recommended the use of glass: 'The young plants may be defended from cold and boisterous windes, yea, frosts, the cold air, and hot sunne, if glasses made for the onely purpose, be set over them.' The growing of plants under glass was developed in Italy, where they had more advanced glass-making techniques: glazed *giardini botanici* were established at Salerno and in the Vatican, though neither has survived.

The development of the orangery as an elegant annex to a noble house, dedicated to growing exotic fruit, has been mentioned in Chapter 67 on oranges. The principle was extended to build pineries, pits under glass for growing pineapples. Other more brutally functional uses of glass for growing plants were being developed at the same time. Charles Louis Bonaparte, nephew of the emperor, built a greenhouse in Leiden to grow medicinal plants from the tropics. Another was set up at the Chelsea Physic Garden by 1681. Greenhouses were established at universities, and were no longer restricted to the rich.

The spread of greenhouses was held back by taxes on glass in England and France, but the taxes were eventually dropped, at the same time that the Industrial Revolution improved glass technology and radically lowered its cost. Glass was for everybody: and that meant that the prosperous middle classes could introduce glass into their gardens. Cucumbers became a more democratic food, though one that still had a little status: you can't grow cucumbers as you grow turnips. The cucumber frame, a moveable wooden frame with a glass lid, became a staple of Victorian gardens. They turn up in *Alice in Wonderland*, when Alice has grown to giant size and fills a small house that belongs to the White Rabbit. The rabbit tries to get in and Alice makes a grab for him: 'She did not get hold of anything, but she heard a little shriek and a fall, and a crash of broken glass, from which she concluded that it was just possible that it had fallen into a cucumber frame.'

Cucumber retained a certain elegance. They were a favoured part of afternoon tea: a light snack that would keep you going without spoiling your appetite for the big meal of the day. Algernon speaks of them in Oscar Wilde's play *The Importance of Being Earnest*: 'And speaking of the science of life, have you got the cucumber sandwiches cut for Lady Bracknell?'

Modern varieties of cucumber are divided roughly into three kinds: known as slicing, pickling and seedless (or burpless) cucumbers. The pickling kind are known as gherkins in the UK, more often just as pickles in North America. The last variety has been developed for people who find the normal slicing cucumbers indigestible. In 2011 cucumbers were withdrawn from sale in Germany, Austria and the Czech Republic after ten people died after being infected by *E. coli*; it was feared this had been introduced in imported cucumbers. Spain, where most of the cucumbers were grown, demanded compensation.

In Spain cucumbers and many other delicate crops are grown on a wide scale under glass. This technology can allow the land to produce a great deal more than it would uncovered, food for an ever-growing human population. The Netherlands has invested heavily in such methods, which are regarded in some circles as the way that agriculture should go: a better management of resources without the same need for potentially damaging chemicals.

The development of polythene from the 1960s to serve the same purpose as glass has made greenhouse farming more accessible. Farming in this way on a large scale is a complex business, involving regulation of temperature, humidity and the movement of air. A greenhouse environment can, as any gardener knows, lead to a build-up of plant pathogens and it is also, obviously, a great deal harder for pollinating insects to operate. Some installations use artificial lighting so that

photosynthesis can take place without the sun; some use additional carbon dioxide. In 2019 the amount of land used globally for greenhouse farming was 1.2 million acres (496,800 hectares). This is a different category from what is called protected farming, in which crops are grown under temporary covers rather than dedicated buildings; in 2019 this added up to 139 million acres (55.6 million hectares).

This kind of farming can ensure something close to a twelve-month growing season for some crops even in northern latitudes. It also requires considerable investment. It makes the agricultural countryside unsuitable for wildlife and human recreation. It is efficient, especially for delicacies like strawberries and, for that matter, cucumbers. Growing food under glass is an increasingly important part of agriculture. It is no longer about extending the growing season: it is about taking the seasons and the weather out of the equation altogether.

Help yourself: Cucumbers with Radishes, *1902, by Max Slevogt (1868-1932).*

SEVENTY

STINGING NETTLE

I like the dust on the nettles, never lost
Except to prove the sweetness of a shower.

Edward Thomas

The digger had finished its work on the dykes that surround my place, leaving a great deal of bare soil. I was thrilled to think what plants would emerge from this fresh start: what long-buried treasures of the seedbank would sprout for the first time in years. I got nettles. I got nettles of exceptional height, vigour and virulence. Stinging nettles are the camp-followers of humanity: where we go, they follow and they prosper. Much as we wish they didn't.

There are six sub-species of *Urtica dioica*, one of which doesn't sting. Nettles are native to Europe, temperate Asia, West and North Africa and have spread all over the world; in Ecuador until 2010 they were used for public punishment of criminals, who were stripped naked for the purpose of urtification. They are found in Canada and in every one of the United States except Hawaii. They are restricted by their need for moist soil; they are less widespread, though still common, in southern Europe. They spread by seed and by rhizomes and stolons (a horizontal creeping stem that puts out roots to make more plants), dying back in the winter and coming again in the spring: often very early in that season. They rise very quickly after fire, giving them a march over their competitors.

Their leaves are deeply serrated and heart-shaped; stems and leaves are both notably hairy. These hairs, or trichomes, define the nettle for us humans. Each hair is a brittle tube connected to a swollen sac of irritating fluid: one that contains histamine and serotonin as well as formic, tartaric and oxalic acids. In other words, the stuff that causes us such discomfort is highly complex, the product of millions of years of evolution. The apparent malevolence of the stinging nettle is no passing whim: it's a defence mechanism that means that they can flourish in fields full of grazing mammals, unattractive to any but a desperate cow or horse. When we brush a nettle, the hairs break off and inject us with this fluid, in the manner of a hypodermic needle. The effect it produces is known as contact urticaria, a form of contact dermatitis. There is, alas, no chemical justification for applying dock leaves to nettle stings to abate their ferocity: you can argue that they work either as a counter-irritant or as a placebo. Or even both.

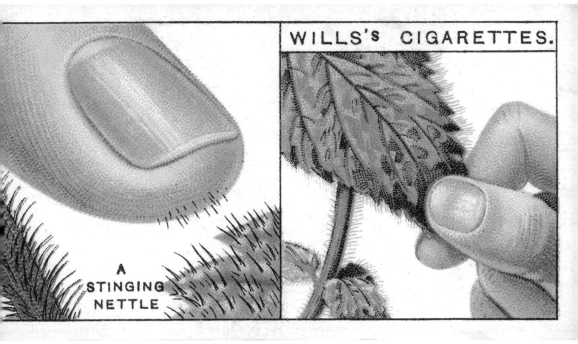

Ouch: illustration for a cigarette card, published by Wills's Cigarettes, early twentieth century.

Nettles have an extraordinary attraction to human settlements: so much so that the presence of unusually dense beds of nettles often indicates long-abandoned human dwellings. Human settlements create ideal and long-lasting conditions in which nettles can flourish: they like highly fertile and preferably disturbed ground, rich in phosphates and nitrogen. Humans have been historically generous in their provision of both, which enter the soil by way of ashes, bones and various kinds of rubbish, the build-up of animal droppings in paddocks, from the disposal of human corpses and, perhaps most significantly, on middens constructed for the disposal of human excrement. Nettles are not what you'd call fastidious: they celebrate all that is least glorious in the human condition and flourish.

The nettles at my place are not profiting from any abandoned settlements. They exult in the run-off of chemical fertilisers from the neighbouring farmland, which reaches us through the soil and along the watercourses, allowing the nettles to flourish at the expense of more picturesque wild flora. This abundance of richness in the soil is found wherever any kind of agriculture takes place, particularly of the modern and intensive kind. It is called eutrophication: too much of a good thing. It's one of those ecological problems that doesn't sound like a bad thing (like many aspects of climate change); conservation organisations find it hard to sell as a major problem. But the phenomenon is damaging to biodiversity.

In Britain you can see the result along roadside verges in agricultural land: a hyperproduction of nettles and cow parsley and very little else. The conservationist Oliver Rackham, in a moment of rage and despair, wrote: 'Almost every rural change since 1945 has extended what is already commonplace at the expense of what is wonderful, or rare, or has meaning.'

It has been speculated that the Romans brought nettles into Britain, but that idea was contradicted when a Bronze Age burial chamber was opened in 2011, and the body inside was wearing a sash made from nettles. Nettles have been used to make clothing for a good 3,000 years: the fibres are strong, though coarse. During the First World War some German Army uniforms and equipment like rucksacks were made from nettles because of a shortage of cotton. Nettles also make a tenacious dye: yellow from the roots, and greeny-yellow from the leaves: you can read an accurate account of the process in Victor Hugo's *Les Misérables*.

Nettles are also a food-plant for humans. They are extremely nutritious, high in vitamins A and C, rich in iron, potassium, manganese and calcium. They are 25 per cent protein by dry weight. Their early arrival in the spring made them the first fresh vegetable annually obtainable; apparently they have a taste between spinach and cucumber. Richard Mabey, in his classic *Food For Free*, recommends a nettle soup with onion, potato, crème fraîche and nutmeg; the chemicals that cause the irritation are destroyed by the process of cooking. Pick your nettles before they have flowered; more mature plants can contain cystoliths that can irritate the urinary tract. The excellent cheese Cornish Yarg comes wrapped in edible nettle leaves. People make tea from nettles, and even beer.

Nettles have often been used to alleviate rheumatism and arthritis and to promote lactation. A Romani treatment for arthritis of the hands is to take a handful of nettle leaves and have a good rub. In one of the fables of Aesop, a boy is stung by a nettle. His mother tells him next time to grasp it boldly 'and it will be as soft as silk'. The dashing and impetuous Hotspur makes the same point in Shakespeare's *Henry IV Part 1*: 'Out of the nettle, danger, we pluck this flower, safety.' A good firm grasp crushes those hypodermic trichomes flat, unable to enter your skin. Apparently.

When we are irritated we are nettled; when the Germans and Dutch get into trouble they say they are sitting in nettles. In the Baltic they say that lightning never strikes a nettle, meaning that bad people get away with stuff. The French warn you not to push grandma in the nettles, a warning not to abuse a situation.

Nettles are valuable food-plants to many non-human species, despite their elaborate defences. They attract froghopper larvae, aphids, psyllids, true bugs, leafhoppers and thrips; beetles, flies and spiders hunt for them in the nettle thickets. Nettles are also a food-plant for the caterpillars of many moths, also for the butterfly species peacock, comma and small tortoiseshell. Every garden should have a nettle patch – as well as a space for more fastidious native wild flowers.

SEVENTY-ONE

CANDIDA ALBICANS

Unaccommodated man is no more than
a poor, bare, fork'd animal.

William Shakespeare, *King Lear*

We may see ourselves as the heirs of a noble race of saints and sages, but to a fungal spore we are much more important. We are an environment: an ecosystem: a place where a fungus can live and thrive and make more of its own kind. Every human being that ever lived is a potential habitat for fungus, and most of us are hosts to fungi at some time or another, often enough for years.

The air is full of fungal spores looking for homes. To many of us that is a cause for rejoicing, since (see Chapter 11 on yeast) these things have given us leavened bread and alcoholic drinks. The payback is that some species of fungi find us the perfect place to live. Perhaps the most unpleasant of these is *Candida albicans*, which makes a comfortable home for itself in and around the vagina, penis and scrotum.

'They settle down in places where they shouldn't,' a medical website explains delicately, though the moral implications of that statement are rather lost on the fungi. My right big toe, when short of nail after being trodden on by a horse, turned out to be perfect for the development of fungi. Few people who live into middle age escape fungal infections around the toenails: ugly and unpleasant but not painful or dangerous.

You can also get fungal infections in the fingernails, though these are rarer. They're associated with people who do a great deal of their work with wet hands: hairdressers, farmers, stable staff. The most well-known fungal infection is the one we call athlete's foot: fungus that grows between the toes. Again this is a result of moisture: sweaty socks, wet floors. The fungus can be passed on when bare feet share wet spaces: showers, swimming pools, sports changing rooms. You can spread it around your body by scratching.

The association of sports with fungal infection continues with the complaint referred to as jock itch. This tends to occur where skin meets skin: most notoriously where thigh and scrotum touch. You're more likely to get this one when young. The same problem can occur in overweight people, who develop folds of skin; women can suffer from the same problem under the breasts.

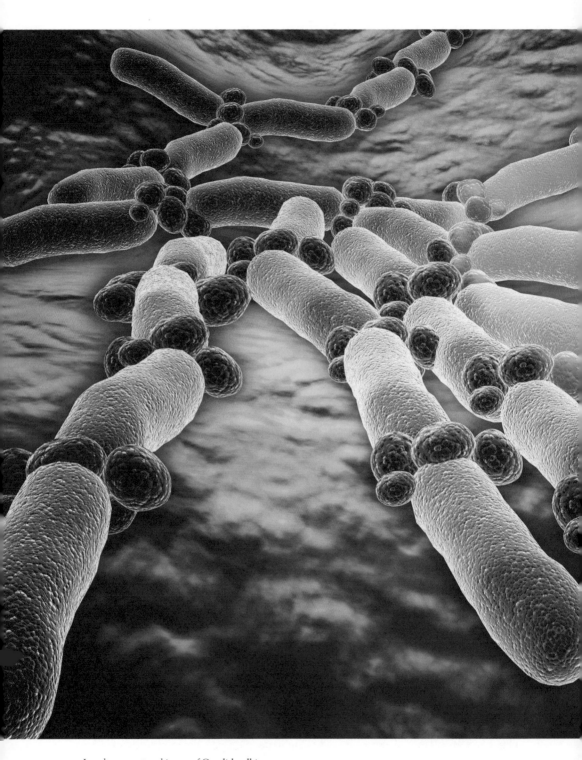

Invader: conceptual image of Candida albicans.

I must confess that I'm feeling the urge to finish this chapter as soon as possible and get on to something nicer, at the same time feeling that this book would be incomplete and therefore dishonest were I to ignore these fungi: and with them the fact that we humans, for all our world-changing technologies, for all the discoveries of science and all the beauties and wonders we have created – for all the fact that we have sent people to the moon, sent spacecraft deep into the solar system, discovered the way life operates, split the atom, built and destroyed cities, cured cancer, produced *Ulysses*, *Starry Night* and the *Goldberg Variations* – are still animals that sweat, that retain moisture in unwanted places, that attract fungi to the most secret parts of our bodies. When we observe that even the greatest people who ever lived have their failings, we tend to say, referencing the Book of Daniel, that they have 'feet of clay'. We may have the minds of angels and the hearts of gods, but we have the feet of athletes.

Fungi in the human body can create greater problems than mere inconvenience. We all breathe fungi in and out with every breath, and this can be a serious matter for people with lungs damaged by previous illnesses, or whose immune system is compromised (from chemotherapy, or from conditions like HIV). The dangerous fungus is *Aspergillus*, a genus with several hundred species, associated with compost heaps, building sites, central heating and air-conditioning. It can be life-threatening to humans.

The list of fungal complaints doesn't stop there. Fungi can affect the ear canals, which is an inconvenience to swimmers and divers. It can affect the mouths of babies. Ringworm has nothing to do with worms: it's a fungal problem that creates red circular sores on arms, legs and scalp. You can get it from touch, from bedding, and from farm animals and pets. The genus of yeast fungi *Candida* can affect the vagina and penis and cause considerable discomfort, along with a white discharge. All these conditions can be treated with creams.

So, if you are still with me, let us hurry on to something more cheering and certainly more comfortable. It is always important to remember that we are humans, and that means that we are a species of mammal. Perhaps it's good that these fungi like to grow on us: they remind us that we should never get above ourselves. As an old saying has it, a certain number of fleas is good for a dog – otherwise it forgets it's a dog.

SEVENTY-TWO
CACAO

Look, there's no metaphysics on earth like chocolates.

Fernando Pessoa

The Dementors make their first appearance in *The Prisoner of Azkaban*, the third of the Harry Potter books. They embody depression, bringing hopelessness and despair. How can you recover from an encounter? 'Professor Lupin was breaking an enormous slab of chocolate into pieces. "Here," he said to Harry, handing him a particularly large piece. "It'll help."'

Chocolate the comforter, the cheerer-upper, the treat, the banisher of misery, the self-awarded prize for finishing another chapter. Anthony Powell, in the war trilogy of his twelve-volume novel *A Dance to the Music of Time*, writes of chocolate. After a dawn start his narrator, Nick Jenkins, second lieutenant, has missed supper and fallen in the river twice at the end of a disastrous day. His company commander gives him a bar of chocolate: 'I had forgotten how good chocolate could be, wondering why I had never eaten more of it before the war. It was like a drug, entirely altering one's point of view. I felt suddenly almost as warmly towards Gwatkin as to Corporal Gwylt…'

Chocolate comes from the seeds of the cacao tree, an understorey plant of moist tropical forest. It's unusual in that it flowers directly on the trunk or the older branches; it's pollinated by tiny flies. It's an evergreen that grows to about 26 feet (8 metres) and is native to the Americas between southern Mexico and the Amazon Basin. The plants develop seed pods, each containing twenty to sixty seeds, usually called beans. These contain a good deal of fat, and also theobromine, a psychoactive substance that enhances mood. In other words, chocolate is not like a drug. It is one.

It was quite literally treasured in the pre-Columbian civilisations of the Americas; the beans were used as currency, and were sometimes forged. They were also used for making a drink. Ceramic vessels dating back to 1900 BC have been found with the residue of cacao. There are continuing disputes as to when the plant was first cultivated. The Aztecs came to dominate Mesoamerica by the fifteenth century AD, and cherished the legend that Quetzalcoatl, the

Opposite Happy all round: 1906 poster advertising Fry's Cocoa.

plumed serpent, was kicked out by the other gods for giving humans the gift of chocolate. The first part of the scientific name *Theobroma cacao* means 'food of the gods'. It was drunk with additions to counteract its bitterness: chilli, vanilla and honey. It was traded and given in tribute: it was at the heart of the civilisation that the Europeans found when they reached America.

Spanish travellers drank chocolate when they met Montezuma at the Aztec capital of Tenochtitlan in 1519. The Europeans liked it and started to add sugar to the drink. They brought it back to Spain and the Spanish court took it up. It then spread across Europe over the next century. The French established plantations in the Caribbean; the Spanish did so in Venezuela and the Philippines.

Turning the bitter beans into chocolate is a hefty process. The beans must be cleaned, roasted and fermented, the shells must then be removed and what's left, the nibs, must then be ground. This is cocoa mass, which is liquefied by heating, to make cocoa liquor. This is cooled and separated into cocoa solids and cocoa butter. To make chocolate as we recognise it, you mix the solids and the butter with sugar; you can substitute vegetable oil for cocoa butter to make something cheaper and less good.

This was a labour-intensive process before the Industrial Revolution. A Dutch chemist, Coenraad van Houten, began to change this. First he added alkaline salts to make the chocolate less bitter, then he invented a press that removed half the fat. In 1847 the English manufacturer Joseph Fry made mouldable chocolate and mass production could begin. In 1875, Daniel Peter of Switzerland added Nestlé powdered milk to invent milk chocolate. Rudolph Lindt, also Swiss, invented a conching machine, which got rid of the gritty texture of early chocolate. John Cadbury established a business in Birmingham in 1824, selling tea, coffee and drinking chocolate. He established a model village called Bourneville on Quaker moral principles of benign capitalism; my grandmother worked there. In 1875 they produced the first Easter egg.

Chocolate had become a fully democratic product. It is remarkable for its ubiquity and for its extraordinary ability to make bad times a little less bad. It's a substance treasured by millions: delivering good cheer and much-needed carbs at a break during a day of physical work, as a treat at the end of a desk lunch, to mark a favour, a welcome to guests in the form of cakes and brownies, for birthdays and other celebrations, as a self-indulgence and as a gift of love. Chocolate is so pervasive in so much of Western society that we scarcely notice its singularity.

These days about 70 per cent of the world's cocoa, the raw material for the manufacture of chocolate, comes from West Africa, about 50 per cent from the Ivory Coast and Ghana, followed by Nigeria and Cameroon. These countries are able to supply the cocoa in quality at cheap prices because the industry uses child labour: up to 2 million children, according to a United States Labor Department report of 2015. Detecting the source of cocoa is not a straightforward matter: it is mostly sold in batches of beans from more than one country. There have been reports that child labour for this work is trafficked between one centre and another: in other words, they are slaves.

The industry has repeatedly been challenged on this subject, and in return has offered deadlines for all kinds of improvements. These were missed in 2005, 2008 and 2010. There was a promise from the industry in 2001 that within four years they would eradicate 'the worst forms of child labour', presumably retaining only the best forms.

There are claims that in more recent times there have been advances with the establishment of schools, support of independent cooperatives and families. But the uncomfortable fact remains that the consuming countries are accustomed to the daily availability of affordable chocolate and there is very little consumer pressure for reform. It is not one of those scandals that makes headlines: it is something we would all rather not know about, since most of us are complicit.

I had promised to reward myself with a piece of chocolate when I finished this chapter, and will do as I said. This particular piece was given to me by my wife after a visit to the local farm shop. It was about twice the price of a bar of Cadbury's. The brand is Tony's Chocolonely. In some circumstances the chumminess of the wrapper message would have irritated me, but not today: 'Hello there. I'm Tony's Chocolonely. I exist to end slavery in the chocolate industry. My mission is to make 100% slave free the norm in chocolate. Together we'll make all chocolate 100% slave free. Are you in?'

SEVENTY-THREE
STRAWBERRY

Strawberry fields forever!

John Lennon and Paul McCartney

More than 2,000 years ago, Ovid set about writing his great work *Metamorphoses*, already mentioned in these pages. He began by recounting the history of the earth in a series of great ages, beginning, of course, with the Golden Age, in which people lived without the cares and troubles of modern life. It was a lost time of perfection in which humans fed on strawberries.

> *Content to enjoy the food that required no painful production*
> *Men simply gathered arbutus fruit and mountain strawberries.*

The arbutus fruit come from the plant known as the strawberry tree, which is unrelated, though its fruit look similar.

Across the centuries strawberries have been associated with pleasure. Also with desire, sex and sin. And tennis, of course. But the strawberry we know today has only been around since the 1750s: it's a hybrid of two New World species.

The strawberry Ovid writes about is *Fragaria vesca*, sometimes called wild or woodland strawberry. It was cultivated in France from about the fourteenth century, as a delicacy that could be grown in gardens. The fruits are very small, though the best have a good flavour. Technically they're not berries but aggregate accessory fruit: that is to say, parts of the flower other than the ovary form parts of the fruit. They grow low to the ground, which was considered by some to make them a lowly product, drawing humans down to the earth instead of up to heaven. No doubt the whiff of sin has always been part of their attraction. Virgil pointed out the ambiguous nature of strawberries in one of his eclogues. Writing about the same time as Ovid, he warned that boys picking strawberries should beware of 'a cold adder lurking in the grass'. The innocent search for pleasure was a dangerous thing.

Pliny the Elder also wrote about strawberries, differentiating between the strawberry and the strawberry tree. These days the latter is classified as *Arbutus unedo*, because Pliny wrote of its fruit, '*unam tantum edo*': I ate only one. This has sometimes been interpreted to mean that the fruit is so wonderful that he could only bear a single one, but he wrote later that the fruit was held 'in no esteem'.

Pleasure principle: detail from The Garden of Earthly Delights, *1490-1500, by Hieronymus Bosch (c.1450-1516).*

Charles V of France was fond of strawberries and caused 12,000 strawberry plants to be established in the royal gardens. Strawberries turn up in Flemish, German and Italian paintings of the late Middle Ages and Renaissance; more of that shortly. They were also thought to have medicinal value and were used to treat depression and colic and to cleanse the blood.

In 1509, Cardinal Thomas Wolsey, chancellor to King Henry VIII of England, staged a banquet in which his guests consumed a new dish: strawberries served with cream. William Butler, a sixteenth-century English physician, said, in words quoted by Izaak Walton in his work of 1653 *The Compleat Angler*: 'Doubtless God could have made a better berry, but doubtless God never did.'

Butler was not discussing the modern strawberry, which he never tasted. This later strawberry is a hybrid, technically *Fragaria ananassa*. It combines *F. virginiana*, from North America, for its flavour, and *F. chiloensis*, which, as the name implies, originated in Chile, for its size. The Chilean plant was brought back to Europe in the eighteenth century by Amédée-François Frézier, a French spy; it's reassuring to note that even then France had a gastronomic intelligence service.

This new variety could be cultivated on a larger scale and soon strawberries were a treat available to all but the poor. The All England Tennis Championships were first held in 1877 at Wimbledon, and from the first, the occasion had a garden party atmosphere. Strawberries became an inseparable part of this. They were an intensely seasonal dish; as the European tennis season developed it came to follow the peak strawberry season as it advances from Madrid to Rome, Paris and finally London. Here strawberries became a summer part of the fashionable afternoon tea. The strawberry tradition continues: 28,000 kilos are consumed annually during Wimbledon fortnight.

Perhaps part of the point is that tennis and strawberries are both associated with sex; after all, tennis was a daring new sport which young men and young women could play unchaperoned. It was from the start a vigorous physical contest played in public by women. Strawberries have been associated with Venus and with female sexuality. Adonis preferred to go hunting rather than spend the day in the arms of Venus, but was fatally gored by a boar. Venus wept and her tears fell as strawberries. In Thomas Hardy's *Tess of the d'Urbervilles*, d'Urberville offers to pop a strawberry into Tess's mouth. '"No – no!" she said quickly, putting her fingers between his hand and her lips. "I would rather take it in my own hand." "Nonsense!" he insisted; and in a slight distress she parted her lips and took it in.'

There are religious implications in strawberries: their tripartite leaves represent the Holy Trinity, the petals of their flowers the five wounds of the passion and the colour of the fruit Christ's blood. They have been associated with the Virgin Mary: but perhaps always with a whiff of danger. The Virgin appears alongside a strawberry bush in *The Little Garden of Paradise*, a fifteenth-century painting by an unknown hand that's now hung in the Städel, Frankfurt. Look closely and you find a little devil lurking among the strawberries.

The greatest strawberry painting is *The Garden of Earthly Delights* by the Flemish painter Hieronymus Bosch. The central panel of the triptych shows a crowd of naked pleasure-seekers, many of them gorging on – perhaps even ravishing – giant strawberries. The painting has nearly as many interpretations as it has romping nudes: perhaps it's what life would have been like if the Fall had never taken place, or perhaps it was a dreadful warning: such behaviour can only lead to the musical hell painted immediately to the right. Either way, the picture is about unrestrained pleasure, and strawberries play a central part.

A neverland of pleasure and secret meanings can be found in the Beatles song 'Strawberry Fields Forever': 'Nothing is real.' It's about the ego loss of LSD (see also Chapter 31 on magic mushrooms). The title comes from the garden at a Salvation Army children's home in Liverpool, which John Lennon knew in his childhood. After his death an area of Central Park was dedicated to his memory and named Strawberry Fields.

When I was a boy it was the custom to make a wish with the first strawberry of summer. These days, with farming under glass and aeroplane transport, strawberries are a twelve-month supermarket staple in the Western world. Many out-of-season strawberries lack flavour and have a woody texture, but we can eat them whenever we want to.

Strawberries turn up in the most famous Zen story of them all, and once again they represent earthy pleasures. A monk, fleeing from a tiger, jumps over a cliff and seizes a vine. The tiger snarls down at him. The monk too looks down: at the bottom of the immense drop another tiger is waiting. As he hangs there, a mouse starts to nibble his vine away. The monk notices fruit growing from the cliff. He takes one and eats it. 'What a delicious strawberry!'

SEVENTY-FOUR
COTTON

Oh Lawdy, pick a bale of cotton!
Oh Lawdy, pick a bale a day!

Traditional song

How many cotton garments are you wearing as you read these pages? How many are at least part-made from cotton? But perhaps you are reading this in bed, and wearing few if any clothes – in which case you are almost certainly lying between cotton sheets, having washed and dried yourself with a cotton towel. When we put fabric against our skin, we choose cotton. Almost always. We like to say that it breathes: it is air-permeable and so it doesn't build up heat and moisture on the inside. It is also, in most forms, soft, adapting easily to body-shape and movement. It's hard to think of life without cotton, whether you are wearing a kikoy to work the fields in Zambia or a white shirt for a business meeting in New York. Cotton is inescapable.

Cotton plants are all species in the *Gossypium* genus; different native species grow wild across the world in tropical and sub-tropical areas. The plant can reach about 2 feet (6 metres) in height, but under cultivation is usually kept down to waist height to make it easier to harvest: at first by hand, in more recent times by machine. The stuff we take from the plants evolved for the dispersal of the seeds. These come packed into a lump called a boll. In natural circumstances this opens to allow the seeds to get blown away by the wind on soft, fluffy threads. Many other, often non-related plants have a similar strategy: the British hedgerow plant old man's beard is a classic example of wind distribution by means of fluff. But if you pluck the bolls before they have opened to let their seeds go, you have all the fluff for yourself.

When did people first realise that these long filaments could be spun together to make a yarn and then woven to make a fabric? It's too far back for rock-solid evidence, but cotton cloth seems to have been invented around the same time in several different places, a process we have seen before in this book, most obviously with the invention of agriculture. There is evidence of cotton fabric in the Indus Valley in Pakistan from about 5500 BC, and in Mexico around the same date. Cotton was already widespread in China as early as the Han Dynasty, 207 BC to AD 220.

It took a while longer to reach Europe, where people relied on wool. This reliance explains the importance of the wool trade: in early medieval times fortunes were made from sheep in a manner incomprehensible to the modern sheep-farmer. It also resulted in the delightful notion that the cotton plant was a tree that produced wool – so it must also grow sheep. John Mandeville wrote about India in the fourteenth century: 'There grows there a wonderful tree which bore tiny lambs on the ends of branches. These branches were so pliable that they bent down to allow the lambs to feed when they were hungry.' He provided an excellent illustration of this remarkable tree. There is an accompanying legend of the Vegetable Lamb of Tartary; illustrations show a sheep growing as part of a plant, attached by its umbilical cord.

Labour-intense: The Cotton Pickers, *1876, by Winslow Homer (1836-1910).*

Perhaps one day European civilisation will ask the question: what have Muslims ever done for us? Among the many things brought into Europe by the Moors when they occupied the Iberian Peninsula and Sicily were paper, the compass, Arabic numbers and mathematics, universal education, widespread literacy, libraries, advances in agriculture and irrigation, and crops that included oranges, apricots – and cotton. Cotton came to Europe with the Moors, who first arrived in the eighth century and stayed till the fifteenth. The advantages of this new fabric were at once obvious.

The snag is that it's hard to grow cotton in Europe. The plant needs a long frost-free period, it needs to be well-watered, it needs a good deal of sun and it requires a period free from rain when the bolls can be harvested. It's a perennial plant, but for cultivation it's grown from scratch every year. It follows that when the European nations began to establish empires and to set up profitable – at least to them – trades, the opportunity to grow cotton in tropical and sub-tropical places was taken up with enthusiasm. The Mughal Empire had been manufacturing cotton cloth in India from the sixteenth century, but the British East India Company was able to take this over and dominate. Cotton did what had seemed impossible: it overtook the spice trade.

The first problem in making cotton fabric is to separate the seeds from the fluff. This is done by the cotton gin, the last word being a contraction of engine. A hand-held cotton gin had been used in India since AD 500, and a roller gin from the sixteenth century. The resulting pure fluff must then be spun into yarn and woven to make a fabric. Cloth manufacture was one of the first things to change under the Industrial Revolution: James Hargreaves invented the spinning jenny and there were other major advances by Richard Arkwright and Samuel Crompton. The crucial step for the cotton industry was made by the American Eli Whitney, who in 1793 invented the modern cotton gin, which drastically speeded up the process of manufacture.

But the picking and baling of cotton was still a labour-intensive process. This becomes less of a problem if you don't pay your workforce, or if you find ways of forcing them to work for very little reward. The people who grew and picked the cotton that clad the British Empire did not grow fat on the deal, as the great Indian freedom campaigner Mahatma Gandhi elaborately and accurately pointed out.

The southeastern United States have an ideal climate for growing cotton, and there the cotton-masters established a system of slaves to make its production economic. The phrase 'cotton is king' summed up an economy based on growing and selling the stuff. This system ended in 1865 after the Civil War, but did not result in instant freedom and prosperity for the ex-slaves. Cotton-growing continues; in 1920 a picking machine was developed to ease the reliance on cheap labour.

We take cotton so completely for granted that we scarcely give it a thought. As I write these words I am certainly wearing at least as many cotton garments as you who are reading them. Cotton is the most widespread non-food crop in the world. It provides income for an estimated 250 million people and employs 7 per cent of all labour in developing nations. And it is extraordinarily damaging. Its growth requires intense use of agrichemicals, in particular pesticides, also fertilisers and herbicides. Areas downstream of cotton farms are polluted by the run-off: places like the Aral Sea in Central Asia, the Indus Delta in Pakistan and the Murray Darling River in Australia.

The production of cotton degrades the soil, so there is a consequent need to move on and establish new areas for plantation, normally by destroying wild habitats. Cotton farming has a very heavy water footprint, in many areas to an unsustainable level. The cotton industry as it stands today is, in the long term, unsustainable. There are initiatives to redress this: to establish responsible cotton farming with fair treatment for those who work in it, by such organisations as Fairtrade. It is a sad thing to think that your underpants are degrading the planet.

SEVENTY-FIVE

SOYBEAN

Animal factories are one more sign of the extent to which our technological capacities have advanced faster than our ethics.

Peter Singer

The supermarket is full of the stuff these days: burgers, sausages, meat pies, meatballs, bacon rashers, escalopes, mince, stewing steak, chorizo, fishcakes, fish goujons, hoisin duck, chicken nuggets… and none of them contains a trace of animal protein. Fishless fish, beefless beef, pigless pig and chickenless chicken are all available: and most of these products get their protein from soybeans. Soybeans are almost preternaturally filled with protein and humans have been cultivating them for human consumption for a good 9,000 years. They are associated with vegetarianism and veganism in the developed nations, to be enjoyed or despised according to taste. Soybeans have become one of the most important crops in the world, and therefore one of the most significant sources of protein. More than 80 per cent of it is fed to livestock.

Soy plants are native to Northeast China, Japan, Korea and Russia; the origin of the domesticated species is probably China, where it was first cultivated. They are comparatively tall, at around 6 feet 6 inches (2 metres), and grow best on good, well-drained soil. They need warmth, and can't abide frost. They were first cultivated in the areas where they grow wild. Like most legumes they take nitrogen from the air and fix it in the soil, and that makes them important in systems of crop rotation. They were so valuable, in terms of protein and in the way they improve the fertility of the soil for the subsequent crop, that in some places they were considered sacred.

The tradition of using soy as a basis for fake meat began in China and is about 2,000 years old. It was a response to the vegetarian tradition of the Buddhist religion, which had recently arrived from India. The tradition continues: visitors to Buddhist temples (often meat-eaters in normal life) will naturally be served vegetarian food, but, perhaps as a politeness to the guests, it will often resemble meat and fish dishes. Buddhist restaurants in China serve elaborate and delicious meals, apparently jumping with dead animals, and all of them made from soy.

Across eastern Asia soy is consumed as bean curd or tofu, not masquerading as meat but as something to be enjoyed for its own virtues. The texture varies from

Fodder crop: tractor in a soybean field, 1943 (artist unknown).

dense and chewy to that of egg custard, and it readily takes up surrounding flavours. Paradoxically, soy is also used to provide the powerful and ubiquitous flavour of much of Chinese cuisine as well as those of other parts of Asia. Soy sauce (or just soy in North America) is made from a fermented soybean paste, roasted grains and brine. Miso, strong-flavoured and proteinaceous, is also made from fermented soybeans. You can use the immature beans in salads; in some varieties the pod is also edible. These young beans are known, a trifle confusingly, as edamame beans: the West has taken to Asian cuisine and uses words taken at random from different oriental languages to describe them.

Soybeans continue to be an important human food in Asia but they were initially something of an exoticism elsewhere: with high protein and low starch they were helpful to those with particular dietary needs, like diabetics. Then came William Morse. Few people have heard of him, but he can genuinely claim to have changed the world. He was the son of a butcher and fresh out of college when he joined the United States Department of Agriculture. He was much encouraged by Dr Charles Piper, who got him to work with soybeans. In 1910 he was growing them at Arlington Experimental Farm, where the Pentagon now stands. The twist was that this time, the soy was being grown specifically as a fodder crop: a crop to feed not humans but their domestic livestock. Piper is quoted as telling Morse: 'Young fellow, these beans are gold for the soil. One must truly stand in awe of their potential power in the life of the western world.'

These words seemed an unambiguous truth when soybeans were planted in the aftermath of the Dust Bowl of the 1930s (see Chapter 35 on kudzu). The planting of soy in these devastated fields returned nitrogen to the soil and played an important part in overcoming this manmade ecological disaster. Soy was grown widely in the United States and fed to livestock. This didn't change farming, but it allowed farming to change. The development of cheap, palatable, high-protein food meant that livestock farmers were no longer dependent on grass for their cattle. They could supplement grazing with other feeds, and so keep more cattle per acre. As the availability of soy grew, pasture could be taken out of the cycle entirely. Cattle could be kept in barns. They need never step outside. Factory farming was now possible.

Pigs can also be fed on soy. Chickens have almost always been fed artificially, rather than letting them survive on what they can forage. Soy, with its elevated protein content, made the feeding and fattening of chickens swifter and more efficient: and with the development of battery farming after the Second World War, chickens fed on soy live their brief six weeks (a chicken can live at least five years in good conditions) sometimes up to twenty birds to a square metre. Soy made possible a radical change to the way we think about livestock: it can be treated more like a plant than an animal, nurtured artificially, treated with chemicals and harvested when ready.

With more and more animals inside, there was an increasing demand for crops to feed them. This led to the development of genetically modified soy: soy has been one of the leaders in the genetic revolution of farming. The most significant advance is a strain resistant to glyphosates: you can apply as much of these herbicides to your soy fields as you like without fear of damaging your crop. Genetically modified soy dominates in the United States and Brazil, the main producing countries; GM products are banned from the European Union. The argument continues: some say that GM crops are the only answer for an increasing human population; others cite the law of unintended consequences and predict ecological disaster (more on this in the next chapter).

The argument in favour of GM soy and the widespread planting of soy is based on the perception that very high meat consumption is essential to human health and happiness. A great deal of rainforest in Brazil has been destroyed for soy farming. In the developed world affordable meat is seen as a basic human right.

The oil from soy is also a useful product, found in margarine, vegan cheese, paint, adhesives, lino and fire extinguishers. There are theories that eating soy is bad for humans: that it can cause breast cancer and the feminisation of men and stunt the growth of children. None of this stands up to close examination: but unquestionably they reinforce atavistic notions of the primacy of meat. The ethical philosopher Peter Singer wrote: 'We are quite literally gambling with the future of our planet for the sake of hamburgers.'

SEVENTY-SIX

ARABIDOPSIS THALIANA

O, what a world of profit and delight
Of power, of honour, and omnipotence
Is promis'd to the studious artisan!

Christopher Marlowe, *Doctor Faustus*

Johannes Thal was a pioneering botanist of the sixteenth century. He did a great deal of his work in the Harz Mountains in northern Germany and put together a work (in Latin, of course, that being the universal language of scholarship) about the plant species he found there. It was published in 1588, five years after his death, and among the many species he described, there is an apparently unremarkable plant now named for its discoverer: *Arabidopsis thaliana* (sometimes called thale cress). For four centuries it was not considered of much use to humanity, but there's now a case for considering it the most important plant that has ever lived.

It's native to Europe, Asia and North Africa, and is most often encountered wild as a random plant of roadside verges and railway lines, dismissed as what we call a weed. It grows up to around 8 inches (20cm) and is a member of the mustard (or cabbage) family. But it's not the relationship with those malodorous vegetables that makes it relevant: it's the way it's related to all other flowering plants.

Well, so are all the other flowering plants (angiosperms), you might say, and with justice. But A. *thaliana* has the advantage of being, in its way, super-ordinary – and more than that, extraordinarily amenable and convenient. That means that you can study it. You can do so far more easily than you could an eight-foot-tall ten-acre stand of maize (corn) or a fully grown redwood or a delicate tropical orchid. The plant was studied in the 1900s, in connection with some early work on chromosomes; mutant strains of A. *thaliana* were studied in the 1940s, but the plant was still of limited relevance. For the brutally practical

Opposite *Model plant: image of* A. thaliana *(right, note earlier scientific name) alongside a tower mustard plant, image from a Swedish flora, 1905.*

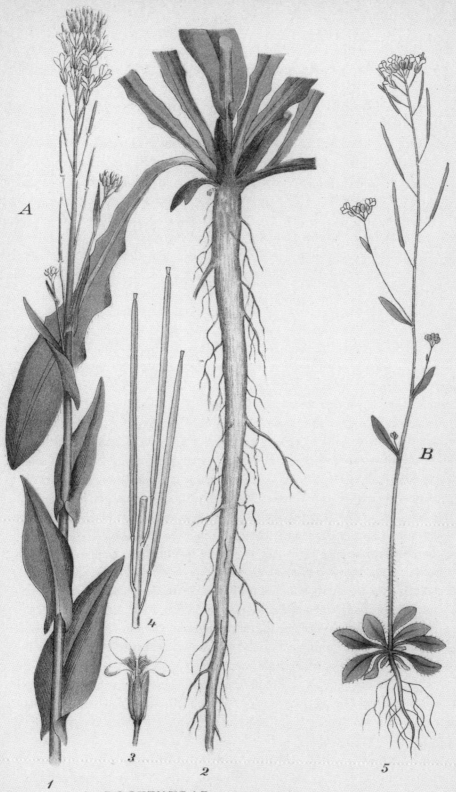

A. ROCKENTRAF, ARABIS GLABRA WEINM.

B. BACKBRÄNNA, ARABIS THALIANA L.

purposes of applied science, A. *thaliana* was considered valueless, because it's not a commercial crop: soy, maize and potatoes were far more important plants to work on.

But in the 1950s and 1960s, the plant began to be used as a model for laboratory studies, from which generalisations about other flowering plants could be learned. The work of John Langridge and George Rédei brought the plant into a wider scientific knowledge, and by the 1980s, A. *thaliana* was the standard lab model.

It was, you might say, about time. Lab work on animals had been continuing through the twentieth century, using favourites that include pigeons, rats and mice. Breakthroughs in animal genetics were made using Drosophila fruit flies, and also the gut bacterium *Escherichia coli* (more widely known as E. *coli*). Since plants provide the oxygen we breathe, all the food we eat, most of our clothes and a quarter of our medicines – not to mention the pages of this book in hard copy – a deeper knowledge of plants was important in both pure research and applied science.

A. *thaliana* is easy to grow: you need no special horticultural talents to bring them from seed to maturity. They are also cheap: not a negligible consideration. They produce plenty of seeds. They are so tolerant they will grow more or less anywhere: as easily in a Petri dish as in a pot, as easily with hydroponics as with compost, as easily under artificial light as in sunlight. They are not only obliging; they are fast: they have a life cycle of six weeks. That means you can go through a lot of generations in a comparatively short time.

That's just what you need if you are studying heredity, genetics and evolution, and such studies are essential to both pure and applied science. With A. *thaliana* in place general principles could be studied with a great deal more precision, with the long-term goal of a total understanding of plant physiology and development: that is to say, how plants work and how plants came to work. A. *thaliana* was the first plant to have its genome completely sequenced, and it's reassuringly and conveniently short. Once a gene had been found in A. *thaliana* it could be more easily found in other plants. So you could, for example, work on the defences of A. *thaliana* against pathogens, and then use what you have learned to make disease-resistant strains. This work can then be carried over into other, commercial species.

A. *thaliana*, then, has become the most important plant in the development of genetically modified crops, and, as said previously, it's a controversial area. The proponents say that it is the only way to feed the human population of the earth: it took humanity 200,000 years to reach 1 billion and another 200 to reach 7 billion; the population stands at 7.9 billion in 2021 and is growing at 80 million a year. One answer to the crisis is to increase the supply of food; some say that GM crops are what the world needs.

The opposition is complex. There are claims that GM crops reduce biodiversity, not only generally but in the crop itself, with fewer local and resistant strains.

There are arguments about the unknown consequences: what, for example, GM pollen will do to the guts of bees, or to what GM plants will do to soil fungi and soil bacteria. There are fears that GM crops will make humans less responsive to antibiotics. Other arguments concern the high cost of the development of such crops. Many GM strains are resistant to the effects of herbicides and pesticides, which mean that more are used. It's claimed that GM crops deplete the soil. There are also political and financial questions: GM crops are patented and the patents belong to the large agribusiness companies; it's been suggested that GM crops serve not the interests of humanity but those of the big companies.

This humble mustard plant bears no responsibility for this; it just grows wherever and whenever it gets a chance. It's a plant that has helped humans to extend their control of the world: just how well we manage this ever-increasing responsibility remains to be seen.

SEVENTY-SEVEN
TULIP

Come tiptoe through the tulips with me

Al Dubin and Joseph A. Burke

I can appreciate a nice tulip if required, preferably the simple uncluttered kind in the Henri Matisse painting *Annelies, White Tulips and Anemones*: a cheering painting of a pretty girl and some flowers. But I can do so without any danger of losing my head. Strange to think that these plants are associated with an event known, even in serious historical books, as tulip mania. This was a period during the seventeenth-century Golden Age of Holland: *tulpenmanie*.

There are perhaps 100 species in the genus *Tulipa*, which belongs in the family of lilies, *Liliaceae*. But that's only a guess: tulips have been cultivated for so long and hybridised so often that there's no clear trail back to ancestral tulips. There's a further complication in that wild tulips are not only very variable from one individual to the next but they also readily mutate and hybridise on their own, and this is confusing to anyone who wants to impose order on the natural world. But it was that combination of malleability and variability that seized the human imagination.

Tulips grow wild in mountains and steppes in temperate parts of western Asia, and they were cultivated in Constantinople (now Istanbul) at least as early as the eleventh century. In the legend of Prince Farhad, the prince was falsely informed of the death of his beloved. He killed himself and tulips sprang from his blood: a strikingly similar story to the legend of Pyramus and Thisbe in Chapter 64 on mulberries. Tulips were prized for their beauty: a beauty that could be cultivated and manipulated under skilled hands.

They didn't reach Europe until the sixteenth century, most likely by way of Ogier Ghislain de Busbecq, who was ambassador in Constantinople to the Habsburgs of Austria and dined with Suleiman the Magnificent. Certainly he claimed to be the tulip-bringer, but they were spread by Carolus Clusius, who established tulips in the Vienna Botanical Gardens and later in the gardens of Leiden University, the *hortus academicus*, demonstrating that tulips could be raised even in more northerly parts of Europe.

One of the odd things about the plant is that they are affected by something called tulip breaking virus. This causes the plant to produce variegated flowers:

Beyond price: The Tulip Folly, *1882, by Jean-Léon Gérôme (1824-1904). A recreation of the Dutch tulip mania of the 17th century – a nobleman guards an exceptional plant while soldiers trample the rest.*

flowers with weird streaks and dramatic combinations of colour. The virus also weakens the plants, so they must be handled with care if producing showy variegated blooms is your aim. Gardeners sought to grow ever-more curious tulips, putting paint powders in the soil, treating the roots with the droppings of pigeons and even borrowing techniques from alchemy. They found they could get just about any colour they wanted apart from pure blue; even the bluest varieties have a violet tinge to them. They have no scent, so they were associated with purity.

The Dutch Golden Age is generally dated from 1588 to 1672: a time when the Dutch maritime trade brought immense prosperity, where a voyage to the spice islands of the Dutch Empire could bring a 400 per cent profit. Holland was the world's foremost economic power, with the highest per capita income. Naturally people wanted something to spend it on: stuff to give them pleasure and show their status.

At this time the Dutch were producing the world's top painters, most notably Rembrandt, Vermeer and Frans Hals. Paintings of flowers became highly fashionable among the rising bourgeoisie, a tradition later adapted by van Gogh (see Chapter 8 on sunflowers). There was also a fashion for the flowers themselves. Especially tulips. Especially funky tulips. People would readily pay big guilders for them.

This gave rise to the speculative frenzy of 1634 to 1637, in which tulips were traded for increasingly drastic prices. Perhaps the most significant thing was the trading at the *beginning* of the growing season: a time when no bulbs were on sale because they were all in the ground. It was the establishment of the first formal market in futures. The Dutch model for trading, with its basis on the stock exchange, is very much the way global business is done today: here was a major advance that came about because of tulips.

The extent of the mania resulted in many tales, or perhaps legends of extravagance: a single bulb traded for a plot of 12 acres of land; single bulbs routinely traded for ten times the annual salary of a skilled artisan; a bulb might change hands ten times in a day, being bulbs not for planting but for buying and selling. There is a story of a sailor who filched such a bulb, unaware of its immense value, and fried it up for breakfast: an unlikely tale as tulip bulbs taste horrible. Different varieties were given increasingly fantastic names – generals and admirals abound. The most expensive of all was Semper Augustus, described thus: 'The colour is white, with carmine on a blue base, and with an unbroken flame right to the top.' Who could resist?

The extent of the madness remains unclear. The story was told and clearly exaggerated in a work of 1841 by Charles Mackay called *Extraordinary Popular Delusions and the Madness of Crowds*. Certainly the market slumped horribly at the end: probably because of an outbreak of bubonic plague in Haarlem that stopped people attending an auction. Tulips remain part of the Dutch tradition, and for a while were Holland's fourth biggest export after gin, herrings and cheese. The Dutch tulip fields are renowned to this day; the largest permanent display is at the Keukenhof in Lisse.

Alexandre Dumas's novel *The Black Tulip* was published in 1850 and touches on tulip mania. Its plot concerns a prize for the first person to grow a black tulip. This is an impossible task, but tulips of deep purple have been bred: Queen of the Night, Black Parrot and Black Hero. A tulip was the symbol of the Iranian Revolution of 1979, in which the Shah was overthrown and the Ayatollah Khomeini took over.

Tulips remain popular as ornamental and as cut flowers. They are often associated with what in Britain is called municipal planting: ultra-formal flowerbeds with all the flowers in straight lines on roundabouts and in front of the town hall. Enthusiasts continue to produce ever more outlandish wonders: Acropolis, Ballade Dream, Cynthia, Big Chief, Aladdin, Blue Heron, Burgundy Lace, Curly Sue, Sensual Touch, Swan's Wing, Belle du Monde, Blushing Beauty. They are all ways in which humans seek to improve on nature: to some eyes the breaking tulip represents a beauty that the wild world will never know.

SEVENTY-EIGHT
COFFEE

I have measured out my life with coffee-spoons

T. S. Eliot, 'The Love Song of J. Alfred Prufrock'

I accepted that coffee was a drink and a mild stimulant, but that wasn't why I drank it. Coffee was also companionship, hope, exploration, flirtation, literature, beauty, truth, friendship, love. In my first year of university it came in the form of Nescafé instant coffee and Marvel dried milk: at least the water was real. By the time I reached the second year the milk and sometimes the coffee were real: perhaps some of the other things too. Coffee enabled us to be together. Would you like to come in for a coffee? It was the drink that opened doors.

How did coffee achieve such massive importance in the world? I like the story of Kaldi, the ninth-century Ethiopian goatherd who noticed that his goats always got lively after eating berries from a certain bush. There's also the story of Sheikh Omar, who was exiled to the desert. He found nothing to eat but these cherry-like berries. They were too bitter, so he roasted them. They were too hard, so he boiled them. The resulting brew revitalised him: he was invited back into society and made a saint.

The real origin of coffee as an important drink for humans is less straightforward, but the species that produces the most popular arabica beans is native to the highlands of Ethiopia and the Boma Plateau in Sudan. It produces a red or purple fruit that's normally called a cherry; this contains two seeds, called beans. In a wild state it grows as a vigorous bush around 10 feet (3 metres) tall. It does best in high elevations where it won't actually freeze. A bush produces fruit three to five years after planting and will carry on fruiting for a good fifty years; 100 is possible. The plants produce a toxic substance to repel predators and this comes in the form of caffeine; the same sort of defence mechanism evolved independently in the cacao, the plant that we use to make chocolate (Chapter 72). There are about 120 species in the genus *Coffea*, and new ones are still being discovered. It's worth adding that about 60 per cent of these species are threatened with extinction in the wild.

Coffee made the short journey across the southern end of the Red Sea from Ethiopia to Yemen through the port of Mocha; it's possible that it was brought there by a Sufi, a member of the Islamic mystical sect. Certainly coffee was taken up in a big way by Yemeni Sufis in the fifteenth century. It helped them to stay

Men's business: Coffee House, 1668 *(artist unknown).*

awake in all-night devotions; the euphoria it produced in high doses was also valued. But soon enough the drink acquired more secular uses: as a pick-me-up, as part of the working day and as a small-scale social event. Coffee reached India; the story is that it was smuggled out by a Sufi named Baba Budan, who strapped the stuff to his chest, to be cultivated in Mysore.

Coffee spread into Europe by way of Venice, which controlled trade in the Eastern Mediterranean. From there the habit of drinking coffee spread: the challenge was to grow it for a growing market. The Dutch East India Company were the first to do this on a large scale, importing coffee grown in the Dutch colonies of Java and Sri Lanka (which was later taken over by the British). In 1637 the English diarist John Evelyn noted that he had drunk coffee, the first British reference in writing to such an experience. It was made by a student at Balliol College in Oxford named Nathaniel Conopios who was later expelled, though not for brewing coffee. The British East India Company started trading in coffee and it became popular in Britain. Coffee drinking spread in Oxford; the Oxford Coffee Club became the Royal Society, now one of the world's leading scientific societies. Lloyd's Coffee House in London became the insurance brokers known as Lloyd's of London.

Coffee needs to be grown in tropical conditions and it was cultivated under the usual polices of colonial exploitation in Asia, Africa, the Caribbean and South America. Costa Rica provided an exception: the population was too small to make large farms viable. Picking by hand required cheap labour. The beans must then be dried, and then roasted and ground. Two species are widely grown, *Coffea arabica* and *C. canephora*, which is usually referred to as robusta. Arabica is sweeter and comprises around 70 per cent of the world's crop; robusta has more caffeine, is cheaper and resists the problem of coffee leaf rust. The degree of roasting affects both flavour and colour; it's all about the caramelisation of the beans. Beans can be soaked or steamed to remove the caffeine.

Coffee reached North America during the colonial period and became popular during the War of Independence, not least because the population wanted to avoid tea, which gave such good profits to the British colonialists. (See the Boston Tea Party in Chapter 54.) Drinking coffee became not just a habit but a patriotic act: today the United States drinks more coffee than any other nation.

Coffee became less popular in Britain in the eighteenth century; tea dominated, not least because it's easier to make. In 1907 the problem was rectified by the invention of instant coffee; the brand Nescafé was from the beginning the most popular. My mother, a garlic pioneer (Chapter 48), had little interest in foreign ideas about coffee and much preferred Nescafé.

In 1938 Achille Gaggia invented the steamless espresso machine, which made it possible to prepare coffee with little trouble, as a small shot or mixed with hot milk as a cappuccino. The machines reached London after the Second World War and inspired a craze: by 1956 there were 300 coffee bars in London. The fashion spread to the United States, most famously at the Caffè Trieste in San Francisco, hangout of the Beat poets. The first Starbucks was opened in Seattle in 1971, named for a mate on board the *Pequod*, the whaling vessel in the novel *Moby-Dick*. There are now 25,000 branches in seventy-five countries.

The popularity of coffee operates on two fronts. It is a drug that stimulates; many people consider it essential for the working day. But perhaps just as importantly it offers an opportunity to stop working. The town of Sloughton in Wisconsin in the United States claims to be the first place to formalise the coffee break in the nineteenth century, and there is an annual festival to commemorate this great leap forward. A coffee break is not only a break but a time to socialise: exchange ideas, discuss procedure, grumble, whinge and flirt. In short, it makes the day better. In some academic scientific establishments, the coffee break is considered a vital part of the process of research: discussion bringing fresh minds to nagging problems.

The most expensive coffee is kopi luwak, made from beans that have passed through the digestive system of a palm civet. This was once collected from the middens of wild civets that lived around coffee plantations; it is now a product of battery farming. It's claimed that the digestive processes of the civet enhance flavour; you can buy the stuff from between US$100 and US$600 for 1 pound (400 grams).

Coffee has a value as a social and business tool. It shows that you are being generous with your time and your resources without the remotest danger of over-committing yourself. 'Let's talk about this over a coffee.' The Nordic countries are the most prodigious per capita consumers: Finland followed by Norway and Iceland; no doubt you need the stimulus in the perpetual night of the Arctic winter. The cultivation of coffee continues to cause ecological and social problems: the crop requires a lot of water, and sun-grown (as opposed to shade-grown) coffee reduces biodiversity. Those that seek a more ethical version of coffee look for endorsement by Fairtrade, Rainforest Alliance or other trustworthy sources.

I drank coffee with immense enthusiasm until I suffered a minor but painful head injury; coffee now gave me blinding headaches so I stopped drinking it. These days I drink rooibos tea from Southern Africa. I saw coffee as an essential… turns out it wasn't.

SEVENTY-NINE
SUGARCANE

Honey,
Ah, sugar, sugar,
You are my candy girl
And you've got me wanting you

'Sugar, Sugar' by Jeff Barry/Andy Kim,
sung by The Archies

Sugar has been described as the substance that nobody needs and everybody craves. It's been claimed that sugar has done more to reshape the world than any ruler, empire or war. It's routinely blamed for a global crisis in nutrition: and yet it's impossible to escape. You'll find sugar in baked beans, ketchup, low-fat yoghurt, spaghetti sauce, granola, soft drinks, flavoured coffee, beer, iced tea, protein bars, health drinks, readymade soup, breakfast cereal, canned fruit, bought smoothies, ready meals, stir-in sauces – apart from all the other foods in which you expect to find sugars like cakes and biscuits and sweets and chocolate bars. You can find sugar in 74 per cent of all food items sold in supermarkets.

But the human taste for sweetness is not to be seen as a recent development. Honey has always been important to the hunter-gatherer lifestyle: it is the most energy-rich food in nature. Research on modern hunter-gatherers has calculated that they get 15 per cent of their calories from honey – and that's not counting the grubs ingested at the same time. This high-calorie, instant-energy nutrition has been described as brain-food: it has been suggested that honey played an important part in making humans what we are.

Sugars can be found in the tissues of most plants, and we take this most obviously in the form of honey and fruit. The only form of sugar not from plants is lactose, which is found in mammalian milk. The substance sucrose is a product of photosynthesis and is found in its most concentrated form in sugarcane and sugar beet; we will move on to beet a little later in this chapter.

Sugarcane grows naturally in Asia and is (like bamboo, Chapter 53) from the grass family *Poaceae*. It has been cultivated for many millennia, probably first in Papua

Opposite *Sweet nature: treatise on sugarcane and baking, from* Tractatus de Herbis *by Dioscorides Pedanius, fifteenth century.*

Bedinia ca· est in tercio gra
du sic in quarto Radix ein
tuiusdam herbe terra cem annos
seruatur Sumitur Decoctione ei
contra farani tussim Auisa· Et
dolor stoi et intestinoz ex bentositate
uel ex fruste· Suppositor ex eo et
trifera magna frigi matricem calefrez
facit 2 mundificat· Salsimentum
ex eo factum uel triseca magna et
rore marino et Ruisse asso et aceto
inretitum meritat

Brina· herba est Nascitur
inter tuticam hieris corrum
pit in uitis psiac in uino habet
acuta et Venenosam· istin mentem
turbat et inebriat si farine reddi in
seantur in re thura caeco 2 mulieres
ex ca fumigentur et Vulua uexaturi
bonapiendi suunt eam· mirta etiam
surfuri cius 2 aceto et ad impetici
nem 2 scrigmenti Aulent corti cum
bino 2 stercore asinino 2 lini semine
cataple fructo ipce suspsliut et scropu
las arto etiam rum radice cortie·
vulneribus ad putrefactie apporti
sinat ea et mundificat

Mulle trassus est cibus
ex duabus causis· ein
nacet spicis ac Rembus· Ana si
insositatem oli· altia per inso
sitatem farine· ig fla moisiz
tenuerit 2 Ansasum· et Venari
via oppilant epre ig a noutates
hic sine eare cum messe manducet
ab ea precuue sine aut epricer:
2 spsionetice

rarisicat est et exti
trarte ig h randu vsar
Mua ca· est meo vrim luidi
hic· in medio· solut dssoluit:

New Guinea, from where it spread across Asia, the cane eaten raw. It was known in Sanskrit as *khanda*; as it spread west it was known in Arabic as *qandi*, hence candy. It spread still further along the Indian trade routes. It was described by Pliny the Elder: 'A kind of honey found in cane, white as gum, and it crunches between the teeth.'

The important breakthrough came in the fifth century when it was discovered that sugar could be stored and transported as granulated crystals: a sweetness that keeps. The Crusaders brought it back with them; it was described as 'sweet salt'. The trade was inevitably controlled through Venice and the Venetians set up sugar plantations near Tyre; by the fifteenth century Venice controlled most of the refining and distribution of sugar throughout Europe and was renowned for its pastries, sweets and sugar sculptures that made a centre-piece for banquets. Sugar plantations were later established on Madeira and the Canary Islands, when these came under European control. A new cane press, developed in 1390, doubled the amount of juice that could be extracted and Madeira dominated the trade to Europe.

Sugarcane needs a tropical or sub-tropical climate, with no danger of frost and a good deal of rain. In these circumstances the plant is staggeringly efficient at photosynthesis, hence the extraordinary quantities of sucrose. Christopher Columbus took it from the Canary Islands to the New World; Hispaniola, now Haiti, had its first sugar harvest in 1501. By the 1520s there were sugar mills on Cuba and Jamaica, and a little later, Brazil. It was a skyrocket of a product with an increasingly eager market. It requires a great deal of labour: the problems of growing, harvesting and processing were solved in the usual way; first with slaves from Africa and then with indentured labourers from the Indian subcontinent. It's been estimated that between 1501 and 1867, 12 million Africans were sent to the New World; not counting the 25 per cent of these that died on the voyage. In 1807 the trading in slaves across the Atlantic was abolished in the British Empire; in 1834 slavery itself was abandoned. The owners, not the slaves, received compensation.

Sugar remained a luxury item: the process of democratisation began with the discovery of beet sugar in Prussia in 1747. Once there were efficient machines to extract the sugar the product could spread. One of the advantages of beet is that there is no itching hurry to get them from field to processing plant: they are quite comfortable lying on the field for weeks after harvest. By 1880 most of the sugar in Europe came from beet, and around this time it was first grown in England, in Lincolnshire. Sugar ceased to be sold in loaves, but, far more conveniently, in bags of granulated sugar.

Since then sugar has become a global commodity, found everywhere in practically everything. It has been compared to fossil fuels: not a vice or a bad habit but something central to the way we have chosen to live. It is inescapable. The use of sugar has been described as an industrial epidemic: the worldwide spread of a non-communicable disease.

There has been a good deal of revisionist demonisation of sugar; I remember a guest in my house fluting to her partner: 'Do you still want sugar? They've only got White Death.' It's associated with the massive global rise of obesity. People's responses to sugar have been described as indistinguishable from addiction.

It remains an oddity that as public health concerns go, sugar is still comparatively low on the list. For many years fat was considered the real villain, and fat-reduced foods were and still are widely available: we have all seen people washing down low-fat yoghurt with Coca-Cola (the full sugar version of Coke contains about nine teaspoons of sugar in a 350ml serving). There has been little urgency to put a tax on sugar, or to provide health warnings on sugar products.

But the fact is that sugar is not bad for you. It provides a lot of readily accessible energy. What's bad for you is taking on more sugar – more of any kind of food – than you burn up. A high-calorie diet is essential to a long day of manual labour; to retain the same sort of diet for a sedentary life leads inevitably to obesity, and obesity is now widely considered normal in developed nations. The solution of excessive sugar intake lies in personal choice.

EIGHTY

HOLLY

Of all the trees that are in the wood
The holly bears the crown

Extract from a traditional carol in Cecil James Sharp's *English Folk-Carols*

Life is hard, but better times are coming. This idea is perhaps the keystone of human existence, the thing that makes life possible. Bad things happen, as we all know. People are capable of great wickedness, as we have seen in far too many chapters of this book, not least the last. But things are getting better: in this great future you can forget your past, so dry your tears…

Such sentiments are, I think, common to every culture and every religion. In the seasonal lands they often take the form of the veneration of spring: every year the time of darkness, cold and lifelessness around the winter solstice, is followed by spring. The darkest time of the year becomes a celebration of hope, anticipating spring and the better life to come. We *will* get through this. These days much of the world celebrates this as Christmas, as we have seen in Chapter 26 on Christmas trees. But the tradition of reverence for evergreen trees at the time of the solstice is still more ancient, and is associated with holly.

As we have seen in chapter after chapter of this book, plant genera often contain far more species than we expect. The genus *Ilex*, or holly, has 568 species, and is the only genus in its family *Aquifoliaceae*. Holly species can be found all over the world, in many different climates and many different forms: deciduous or evergreen, as trees, shrubs or climbers, from sea level to 6,600 feet (2,000 metres). They produce small fruits, technically drupes, that can be red, brown or black, occasionally yellow or green; each contains up to ten seeds.

But one species dominates this sprawling genus, at least in terms of human apprehension and human history. This is the European holly *I. aquifolium*, more often called just holly, as if it was the only species in the genus that matters. At the time of the December solstice in the woods of northern Europe it seems to be the only plant in any genus that matters: on cold grey days with grey leafless vegetation the holly is an explosion of colour and life: lustrous green leaves and berries that glow bright red. Here is obvious life at the most lifeless time of the year: who could fail to rejoice in it?

Opposite Mighty symbols: a holly bush plays a part in the La Dame à la licorne tapestry, late fifteenth century.

The frost softens the berries and makes them more palatable to birds, who eat the fruit and pass the seeds out in due course, ready for germination in the spring; they are an important winter food for many species. They are poisonous to humans, causing vomiting and diarrhoea: it's been calculated that as few as twenty berries can be fatal to small children. The bitter taste of the berries ensures that they are not a food of choice. The trees grow well, often in the understory of woodland, and their year-long prickly exterior creates well-protected hiding places for small birds. This protectiveness appeals to humans as well: holly traditionally offers protection against evil spirits.

It is impossible to know how long holly has been venerated, but the practice surely predates Christianity. There are lost traditions in which a boy dressed in holly and a girl dressed in ivy paraded through the village at the solstice, two evergreens to ensure the continuation of life. Ivy was considered feminine because it needs something to cling to. In Celtic mythology the Holly King rules the world from the summer solstice until the winter solstice, but is then conquered by the Oak King, who rules until the summer solstice comes round again. The Holly King is often depicted as a giant clad in holly leaves, carrying a holly club. Both these figures seem to be united in the Green Man, an ancient and ambiguous figure full of life-force: you can find his face in the ceiling bosses of Norwich cathedral: fierce but not actively malevolent and covered in leaves. Are they oak or holly? The leaves of the two plants are superficially similar. In the great poem *Sir Gawain and the Green Knight*, the Green Knight, who is a lot like the Green Man, enters the court of King Arthur bearing holly as a sign of his peaceful intentions, for holly is about hope:

> But in one hand a solitary branch of holly
> That shows greener when the groves are leafless.

Hollies in hedges were often left to grow to their full height, standing proud of the other plants in the barrier. It's been suggested that this was to block the paths of witches, well known for using hedge-tops as highways; it's also possible that such occasional markers were useful sightlines for ploughmen trying to keep their furrows straight. Hollies were planted to protect nearby dwellings, not just from evil spirits but also from lightning; they've been associated with the thunder god Thor. It was considered bad luck to fell a holly, but there were no prohibitions on cutting boughs, or on coppicing them. (To coppice a tree you cut it down, leaving a living trunk from which several new shoots will grow, giving a multi-trunked tree.) Holly wood is good and close-grained, used for knife handles, mathematical instruments and engraving. It is white and accepts dye readily. It has often been used for doorsills: again as protection. Seasoned holly makes a good firewood that burns hot: charcoal made from holly was often used in the making of swords and axe-heads.

Like many other customs, holly has been taken more or less seamlessly into Christianity: the berries became a symbol of Christ's blood, the spiky leaves the crown of thorns; an old name for holly is 'Christ's thorn'. This is most widely known from the nineteenth-century carol 'The Holly and the Ivy', but it comes from far older traditions:

The holly bears a berry
As red as any blood
And Mary bore sweet Jesus Christ
For to do poor sinners good!

Holly has been cultivated for its decorative and symbolic qualities. The European species was introduced to North America for those reasons and is now established as an unwanted exotic in forests along the west coast of America from California to British Columbia. Elsewhere, other species of holly are under threat from habitat destruction: one has gone extinct and 100 more are threatened with extinction.

But let us leave the last word with Shakespeare, from a song in *As You Like It*:

Heigho! sing heigho! unto the green holly:
Most friendship is feigning, most loving mere folly.
Then heigho, the holly!
This life is most jolly.

EIGHTY-ONE
DRY ROT

I'll huff and I'll puff and I'll blow your house in.

Extract from traditional story, The Three Little Pigs

When you're trying to conquer the world, it can be vexing when your plans are compromised by a little fungus – but it's happened time and again. It's equally troubling when buildings made to shelter the great are brought down by the same fungus: small, subtle, slow and untiring. HMS *Queen Charlotte* was categorised as a first-rate ship of the line: the best a Royal Navy can get. This ship was launched in 1810, and carried 104 guns: a fearsome weapon of war, Lord Exmouth's flagship at the bombardment of Algiers in 1816, famous for discharging a black sailor named William Brown because she was a woman. The more shocking truth of the *Queen Charlotte* is that the ship was riddled with fungus; in the first six years of existence the navy had already spent more than the cost of building it on subsequent repairs.

Fungi, as stated before, are not plants. That means they must eat. Fungi tend to feed on improbable things: many species subsist on dead and/or decaying matter. That makes them essential to the natural ecosystems they inhabit: digesting, for example, the leaf litter of the forest floor and recycling the nutrients in a manner that makes them available to other life forms. It is the action of fungi that makes rainforests possible. The way they feed is important and beneficial to us humans on many levels, as we have seen not least in Chapter 11 on yeast and Chapter 21 on *Penicillium*. But there are times when the life of fungi is seriously inconvenient to the lives of humans. We have looked at this in Chapter 71 with fungi that use the bodies of living humans as habitat; there are also fungi that compromise the structures that human create.

We have celebrated the use of wood as a material for construction in Chapter 9 on oak, Chapter 62 on teak and a good few others. There are species of fungi that eat wood, and that causes problems when humans make things out of wood. The fungi that cause dry rot in buildings and in ships feed on the cellulose in wood, and that's the substance that binds the fibres of wood together. Infected timber often looks sound for as long as moisture holds the decayed wood in place, but as it dries

Opposite Eater of houses and ships: dry rot fungus, image from Atlas des Champignons Comestibles et Vénéneux, *1891, by Léon Dufour (1780-1865).*

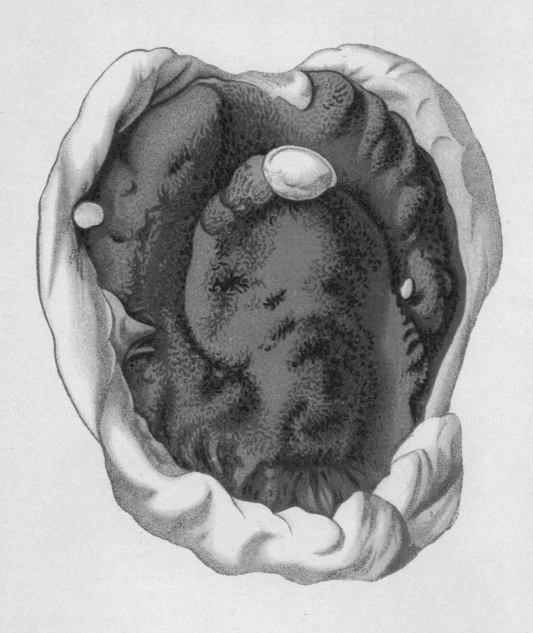

Nº 141. Merulius lacrymans. *Mérule pleureur.*

the truth is revealed and the wood crumbles into dust. It's the most dismaying conjuring trick: the home that looked as if it would last forever and certainly last you a lifetime is crumbling all around you: your best hopes turned to dust.

The term dry rot is used promiscuously to mean both the symptoms of infestation and the fungi themselves. There is also a confusion of the terms dry rot and wet rot; it can be argued that the term dry rot is in itself a nonsense, because the fungi can't get active without fairly high levels of moisture. Mostly we are referring to the species *Serpula lacrymans* in Northern Europe and *Meruliporia incrassata* in North America.

It's likely that the term dry rot came from fungal growth in wooden ships. The condition was often unnoticed until a ship was hauled out of water for repairs – but it was there all along; it wasn't a product of the new dryness. The wood collapses when it dries, revealing the extent of the devastation. Samuel Pepys had to confront the problem as inspector of the British Navy, and he wrote his findings up in a report to the Admiralty Board in 1684: 'The greatest part of these 36 ships… were left to sink into such distress, through Decays contracted… that several of them… lye in danger of sinking at their very moorings… the planks were in many places persish'd to powder.'

Dry rot was a serious problem to everyone who was ever involved with wooden ships, and they were the only kind available until ironclad warships were first constructed in the nineteenth century in Britain and the United States. The *Mayflower* set sail with its pilgrims to the New World in 1620, but it had to turn back twice. That was because its companion ship, *Speedwell*, was leaking, riddled with dry rot. Eventually the *Speedwell's* voyage was abandoned: would the ship's passengers have changed the history of America and the world? But we are never to be told what would have happened.

The fungus gets into the wood by means of spores. These can stay dormant in timber at low humidity for decades, occasionally visible as a fine orange dust. In order to germinate and thrive they need the right conditions, and these can readily be supplied: increased humidity and poor ventilation. That releases the fungi – in the form of the threads called mycelia – to grow through the wood, feeding as they go. Seasoned timber has a moisture content of around 14 per cent; the water content needs to be close to 28 per cent for germination to take place – but once that has happened the fungi can cope with humidity as low as 20 per cent, so dry rot can be looked on as a continuing process of wet rot. When unseasoned wood was used in the construction of ships or houses, it was an invitation to the fungus to step in.

Fungus needs access to air: you have nothing to worry about for as long as your wood is submerged. Venice is built on sunken piles of wood; some of these have lasted 1,000 years because they're not exposed. Problems come when the air and the moisture come together: dry rot was a great destroyer of the railway sleepers (ties); in the United States the fungus *Lentinus lepideus* was known as the train-wrecker.

The fungal species that cause the damage are naturally demonised, as if they were actively malevolent; 'aggressive' and 'property-killer' are terms used by people who make a business of getting rid of the stuff. This can be a difficult process, one that uses ethylene glycol, more familiar as antifreeze, which is poisonous and polluting. Drastic remedies are often required. Once you've got rid of it you then have the problem of stopping it coming back: reducing humidity, controlling condensation and heating evenly will do the job. Fungal growth can also damage electrical and electronic equipment, leather, paper, textiles and optical equipment; there are even species that can extract mineral nutrient from glass.

Throughout history dry rot and other forms of fungi have infiltrated some of the most grandiose constructions of humanity. Their effect, small, subtle and cumulative, has compromised our ambitions across the centuries, reminding us that our divorce from nature is not and never can be declared absolute.

EIGHTY-TWO
OLIVE

And the dove came to him in the evening; and lo, in her mouth
was an olive leaf pluckt off: so Noah knew that the waters were
abated from off the earth.

Genesis 8:11

Which god should have the honour of becoming patron of Athens? It was decided that Athena and Poseidon should have a gift-off, and Cecrops, king of Athens, would decide which god came up with the better offer. Poseidon struck the ground with his trident and caused a salt spring to emerge – undrinkable yes, but offering access to the sea and therefore power. Athena offered the first domestic olive tree: wood, oil, food, light and luxury: in short, peace and prosperity. Cecrops accepted her gift, the city was named in her honour and the olive was at once central to civilisation.

The olive tree is a symbol of peace to this day. In the Book of Genesis, Noah let a dove fly from the ark: it returned with a sprig of olive in its beak to show that the flood was abating, there was land above the water, the horrors were over and a new and better life could now begin. I had a spat with an old but tactless friend; he made a small gesture of appeasement. I wrote: 'If that is intended as an olive-branch, consider it accepted.' The plant goes very deep into Western hearts and minds.

The olive tree is a member of the family *Oleaceae*, along with lilac, jasmine, forsythia and ash, perhaps a surprising mixture. Olives grow as evergreen trees or shrubs, seldom higher than 50 feet (15 metres). They are long-lived, often measuring their lives in thousands; there are sound claims for trees 3,500 years old and wilder assertions for still older trees. They look their age, too, growing into gnarled and fantastic shapes, perhaps best admired, like so many plants, in the paintings by van Gogh. They have small white flowers and produce berries – drupes – with a single stone or pit.

They are among the oldest of all cultivated trees but the mystery is that humans ever discovered that the fruit was palatable, and that it creates the most sublime oil. As everyone who has tried eating an olive from the tree knows, they taste foul.

Opposite Noah's ark: the return of the dove with the olive sprig; Catalan book illustration c.AD 970.

columbe

noe

They must be cured and fermented before being eaten, a process that requires a few days in lye (obtained from wood ashes) or a few months in salt or brine. This removes the oleuropein, a bitter compound. (We tried this at home with modest success with fruit from plants grown in pots under cover in England.)

The oil is obtained by crushing the fruit, traditionally with a millstone, more recently between steel drums. The liquid obtained from the pressed fruit is still very watery. Oil and water were traditionally separated by gravity – oil is lighter – but is now done more quickly and efficiently in a centrifuge. Oil produced by purely mechanical means is virgin oil; if it passes criteria of low free acidity and absence of certain defects, it is extra virgin. Lower-grade processed oils are refined with heat and/or chemicals.

It's no exaggeration to say that the oil obtained is venerated. It plays a significant part in Roman Catholic, Orthodox and Anglican versions of Christianity: in the ordination of priests, in the anointing of the sick in the last sacrament, in baptism, and in some nations the coronation of kings and queens. The lamps in Orthodox churches are fuelled with olive oil. Olive oil is traditionally used for the Menorah, the eight-flamed lamp that is the centre of the eight-day Jewish festival of Chanukah. The prophet Mohammed recommended massage with olive oil: 'It is a blessed tree.'

Olive oil also had a cosmetic use: bodies were anointed with oil to look and feel more beautiful; body-builders still do this, though they mostly use baby oil, a by-product of the petroleum industry. The handmaids of Circe bathed Odysseus in *The Odyssey*:

> *The bathing finished, rubbing me sleek with oil.*
> *Throwing warm fleece and a shirt around my shoulder*
> *She led me in…*

Pliny the Elder, a man with a taste for austerity, wrote: 'The natural properties of olive oil provide the body with warmth and protect it against the cold; it cools the head when hot. The Greeks, progenitors of all vices, have diverted the use of olive oil to serve the ends of luxury by making it available in gymnasia.'

Olives were first cultivated about 7,000 years ago, above all for the oil, which was used for cooking and for light, for ornament and for religious ceremonies. The tree probably spread westward from Mesopotamia, though some prefer Egypt. Olives were taken up in a big way by the Minoan civilisation of Crete; it's likely that Minoan power came from their trade in olives. It was believed that olive trees couldn't grow more than a certain distance from the sea: Theophrastus, the Greek philosopher and botanist of the fourth century BC, suggested a distance of around

Opposite Symbol of hope: Angel with Olive Branch, c. 1475-80, by Hans Memling (1430-94).

30 miles (50 kilometres) as the maximum distance from the sea at which an olive could thrive. He also gave clear instructions on grafting and vegetative propagation of olives. Olives were considered part of the trinity of Greek staples: wheat, olive and grape.

The olive retained its extraordinary significance into the Christian era. Jesus was betrayed in the Garden of Gethsemane: a word that means olive-press. He ascended into heaven from the Mount of Olives (Mount Olivet) outside Jerusalem.

Today olives are grown all round the Mediterranean – Greece still has the highest per capita consumption – and in many other places with a Mediterranean-type climate: South America, South Africa, China, Australia (the olive has become an invasive species in South Australia), New Zealand, Mexico and the United States. Spanish missionaries brought olive trees to California in the eighteenth century. Oil makes up 90 per cent of the world's olive crop; the remaining 10 per cent is table olives. Fruit for both oil and table can be picked at different stages of ripeness for different flavours: green and unripe, semi-ripe and black or fully ripe. Timing is important: oil made from green olives will give you a more bitter taste, but over-ripe olives produce a rancid oil.

Olive is an ancient crop that mostly comes from ancient trees, but there are modern problems and threats for the farmers. These include olive quick decline syndrome. This stems from a bacterium, *Xylella fastidiosa*, which has caused devastating damage to olives in southern Italy. The Mediterranean diet, with its heavy use of olive oil, is considered to reduce heart disease and obesity.

The olive has remained important as a symbol into more secular times. The Great Seal of the United States, a design finalised in 1782, shows a bald eagle with an olive branch in the talons of one foot and a bunch of arrows in the other, telling the world that the nation has 'a strong desire for peace but will always be ready for war'. The badge of the United Nations, adopted in 1946, shows a map of the world within an olive wreath. In 1949 the first International Peace Conference was held in Paris and adopted as symbol Picasso's *Dove of Peace*: a beautifully simple outline of a dove carrying a sprig of olive.

EIGHTY-THREE
RAFFLESIA

It is perhaps the largest and most
magnificent flower in the world.

Sir Stamford Raffles

As soon as I was old enough to take the bus by myself – I was nine, but these were easier times – I used to travel every Saturday from our home in Streatham in South London to the Natural History Museum in South Kensington. I would ascend the steps of this great treasure-house – every ornate brick proclaimed it as such – and every time as I entered the Central Hall the first exhibit I set eyes on was a plant: the most monstrous flower on the planet, more than 3 feet (1 metre) in diameter and weighing 24 pounds (11kg). This chapter and this plant are about the human sense of wonder at the natural world.

The flower was first discovered by a Frenchman, Louis Auguste Deschamps, but he got no credit for it. The genus *Rafflesia* is named for Sir Stamford Raffles, the British colonial governor who was to found Singapore. It contains forty species; the monster of South Kensington is *Rafflesia arnoldii*, named for the British discoverer who was a good few years behind Deschamps. *Rafflesia* is the common name of this species.

Deschamps set sail in 1791 on *l'Espérance*: an optimistic name for a ship that set out on one of the unluckiest voyages since that of Odysseus. The principal aim was to seek a ship called *La Perouse*, which had vanished somewhere in Oceania. They failed in this endeavour. They had terrible weather and were affected by disease: of the 119 crew, eighty-nine died. The ship was then seized by the Dutch and taken to a Dutch colony on Sumatra; Deschamps stayed there until 1802. He was a surgeon with botanical interests and he explored the rainforests with some enthusiasm. He took specimens and copious notes about *Rafflesia*.

He then sailed back to France, but his ship was seized by the British, who were at war with France at the time, and his papers and specimens were confiscated. These were lost for a considerable time, but turned up for sale in 1860, and were bought up by what was then called the British Museum of Natural History. Once there they were lost again, and they weren't found until 1954. It's been suggested that they were not so much lost as hidden: the glory of discovering, describing and naming the biggest flower of them all should go to Britain.

Floral giant: nineteenth-century image of Rafflesia.

Rafflesia is not only a plant with a big flower, it is a very rum plant indeed. It has no leaves or stem, and doesn't trouble to photosynthesise. It is a parasite of vines in the genus *Tetrastigma*, and exists mostly as threads within the host. The vines are climbers, growing on understorey plants in rainforest, with a taste for the introduced *Coffea canephora*, which the alert reader will recognise as the source of robusta coffee (see Chapter 78). The plant only becomes visible when the vine sports a bud that's not of its own making. This can take a year to develop, but at the climax of the procedure it opens into the most colossal flower – which is all that can be seen of the plant. It is the colour of steak that's been left out in the sun and it smells of rotting meat, duping insects with a taste for putrefaction. They enter the vast flowers in search of a meal and inadvertently pollinate them as they leave disappointed, only to be fooled again by the next overwhelming flower.

There are flowers that appear to be bigger than *Rafflesia*, most notably the titan arum, which can be 10 feet (3 metres) tall, but this is not a single flower; it is a collection of flowering parts and is therefore not a flower but an inflorescence. The tallpot palm flowers once in its life after sixty years of growth, and its branching inflorescence can be as much as 26 feet (8 metres) long; it comprises several million individual flowers.

That makes *Rafflesia* the champion, admittedly in a slightly pedantic way. Either way its glory has been vastly appreciated. In 1818 Joseph Arnold was collecting plants on Sumatra, in a British colony there named Bencoolen, now Bengkulu; Raffles was the Lieutenant-Governor. Arnold wrote: 'Here I rejoice to tell you I happened to meet with what I consider the greatest prodigy of the vegetable world. I had ventured some way from the party when one of the Malay servants came running to me... to tell the truth, had I been alone, and had there been no witnesses, I should, I think, have been fearful of mentioning the dimensions of this flower, so much does it exceed every flower I have seen or heard of.'

Arnold died soon after this discovery; his coloured drawing was completed by Lady Raffles. Two species of *Rafflesia* were described at a meeting of the Linnean Society in London in 1820. This was all done in rather a hurry, in case Deschamps – now working in a hospital in northern France – claimed his fully justified priority. The plant was officially described in 1821 and given a scientific name to glorify Britain: name-checking both Raffles and Arnold and keeping the French well out of it.

The plant is found on Sumatra and Borneo. It has been traditionally used to aid both delivery and recovery in childbirth, and also as an aphrodisiac, perhaps a response to the nature of the bud, not a point that needs labouring. A wax model was obtained by the Royal Botanic Gardens at Kew and put on display; in 1877 a poet called J. Hunt Cooke came up with these lines:

> *What strange gigantic flower is here*
> *That shows its lonesome pallid grace*
> *Where neither stems nor leaves appear?*

It is indeed a prodigious thing. In the course of its history it has made humans look somewhat small-minded, but the flower rises above such things in all its stinking majesty.

EIGHTY-FOUR
TOMATO

For goodness' sake, don't touch me tomato

Tomato, a calypso by Sam Manning

It's hard to have a meal without tomatoes cropping up in some form or other: ladled over pasta, spread over pizza, adding richness to curries and freshness to stir-fries, in deep red sauces with meat, in stews and broths, in hot and cold soups, in salads and garnishes, to go with cheese in a sandwich, as a sauce with baked beans, as a salsa, giving sweetness and light to chilli dishes and thrown over everything in the form of ketchup. Difficult to contemplate life without tomatoes: and yet they were treated with deep suspicion in many cultures long after they became available as part of the Columbian exchange. It was centuries before tomatoes were widely accepted; in the United States they weren't considered popular food until the early 1900s.

Tomatoes are technically fruit but legally they became vegetables in the USA after a court case of 1893. In 1887 a tariff was imposed on vegetables but not on fruit. In the case of Nix v Hedden it was claimed that the tomato was not dutiable because it is a fruit – as indeed it is, it's an edible berry. Other fruit we routinely eat as a main course rather than a pudding include bell peppers, cucumber, green beans, aubergine, avocado and squash. But never mind science: the US Supreme Court ruled that tomatoes were vegetables because you eat them with the meat course.

This is a crisis of nomenclature: a clash between technical and vernacular use of words. In loose, uncircumscribed use of language, dogs are animals but chickens are not: both are members of the animal kingdom, but the word animal is often used informally to mean mammals of the non-human kind – and by the same token, apples, like everything else we call fruit, are technically vegetables, belonging to the plant or vegetable kingdom, while aubergines are technically fruit... and the law makes itself up as it goes along for the usual reasons.

The species *Solanum lycopersicum* is native to Mexico and Central and South America: the word tomato comes from Nahuatl, the Aztec language word *tomatl*. (Other English words derived from Aztec include avocado, chocolate, chilli, coyote, mescaline, peyote, ocelot and shack.) It grows as a vine, on weak stems that need support in cultivation, and generally reaches between 3 and 10 feet

Favourite food: boy in a tomato field, 1945 (artist unknown).

(1 and 3 metres) in height. Tomatoes are native to tropical highlands; wild fruit are the size of peas and the hairy stems form roots when they come in contact with the soil. Along with potatoes and aubergines, they are members of the nightshade family, and the toxin tomatine can be found in leaves and green fruit. Tomatoes were cultivated in Mexico from about 5,000 BC, and perhaps elsewhere.

The Spanish conquistadors found out about tomatoes from the Aztecs. Bernardino de Sahagún, a Franciscan missionary (see also Chapter 31 on magic mushrooms), wrote: 'There were sellers of different varieties of tomato in the markets. Those who dealt in tomatoes usually sold the large and the small ones that were yellowish-red when they were at full ripeness.' It's likely that Hernán Cortés was the first to bring the plant back to Europe after his capture of Tenochtitlan, now Mexico City. Once there the plant was grown more for its ornamental qualities than for food.

The earliest mention of tomatoes in European literature dates to 1544 and is from Pietro Andrea Mattioli, who combined the work of physician and botanist in time-honoured fashion: 'Another species has been brought into Italy in our time, flattened like red apples and composed of segments, green at first and of a golden colour when ripe.' The golden colour is intriguing: the fruit were named in Italian *pomi d'oro*, golden apples; in modern Italian *pomodori*. Red is now standard; golden tomatoes are a mild exoticism.

Tomatoes eventually made the transition from ornament and exoticism to staple food, but it was a slow business. They reached India and China, first via Portuguese traders and their colonies in Goa and Macao. But tomatoes were never obvious peasant food: they are not filling, they work best as an addition to filling and sustaining food, so they are – at least on the subsistence level – a luxury. The toxic potential of tomatoes was not exactly a recommendation either – and perhaps there was also a problem that the tomatoes didn't look like familiar food, being brightly coloured and – quite obviously – a fruit, even if one without much sugar in it.

The first European recipe book that mentions tomatoes is Italian and dates to the late seventeenth century. It recommends that tomato be used 'in the Spanish style', so the implication is that the Spanish led the way. Bit by bit, tomatoes were accepted as an easy way of giving a lift to staple foods, adding savour to filling grain-based items and proteinaceous meats. Perhaps the classic example is *spaghetti pomodoro*: pasta made from wheat, served with a simple tomato sauce, flavoured with garlic and herbs.

The origins of tomato ketchup are obscure, as is the origin of the word itself. I prefer the idea that it comes from the Cantonese term for tomato, *fan qie tsap*, which means literally 'foreign aubergine' (*fan qie zhi* is tomato sauce). Ketchup was sold in bottles in the United States in 1836, and by F. J. Heinz in 1876. Modern ketchups have a hugely increased sugar content: there's a teaspoon of sugar in every tablespoon of ketchup. The vinegar content makes it a sweet and sour sauce, but the sweetness dominates.

As tomatoes gained not so much popularity as ubiquity, entrepreneurs sought the best way of getting tomatoes to consumers. One of these was the development of the so-called 'square tomato' at the University of California Davis. They were not exactly square or, for that matter, cuboid, but they were less likely to roll off a conveyer belt. This development, along with the tomato harvester, made it easier to gather and can the tomatoes.

Perceived consumer demands have led to the development of beautiful shiny red tomatoes with little flavour and low sugar content. Greenhouse cultivation of tomatoes now takes place in many climates, with cultured bumblebees doing the job of pollination. There are claims that lycopene, present in tomatoes, helps to prevent cancer and cardiovascular disorders, but these cannot legally be made when selling them.

How do you pronounce the damn word anyway? These days standard British English is 'tom-ah-to', and standard American 'to-may-to'. In America before the Second World War, British pronunciation was generally favoured by the upper-class members of this classless society. This gave rise to the song by George and Ira Gershwin, 'Let's Call the Whole Thing Off', in the 1937 musical *Shall We Dance* starring Fred Astaire and Ginger Rogers. Before the couple dance away on roller skates they sing:

You like tomayto
And I like tomahto...

EIGHTY-FIVE
ALMOND

The keeping of bees is like the direction of sunbeams.

Henry David Thoreau

This may seem a slightly odd choice. Almonds are nice but they're hardly central to our lives. They're a highly acceptable bonus to some and an important health-food to a good few, but at first glance they don't seem to be an important part in the making of human history. We encounter them in the marzipan that lies between the fruit and the icing in a Christmas cake, or in fancy biscuits, especially macaroons and macarons, or in the one-sip-is-enough liqueur Amaretto. But almonds are at the heart of one of the most remarkable annual events in the great interaction between humans and plants that shapes the landscape of the planet.

It takes place in Central Valley in California, where 80 per cent of the world's almonds are grown. The place is ideal for them: snow on distant mountains melts beneath the summer sun and flows into the valley throughout the summer to irrigate the trees. In February they all come into flower together: 1 million acres (0.4 million hectares) of them. For two weeks the valley blossoms: there are 20,000 flowers on every tree and 90 million trees, so that makes 2.5 trillion flowers. Van Gogh painted the best pictures of almond blossom; the most famous celebrates the birth of his nephew, named Vincent in his honour.

There is a snag to the blooming magnificence of California. If the trees are to produce almonds, they must be pollinated. The male grains of pollen must find their way on the female stigma in the flowers, and the job must be done by insects. And that's a problem, because there aren't any. Or very few. This is a monoculture: nothing that is not an almond is tolerated. The poisons that kill damaging insects also kill the helpful ones; the poisons that kill all plants that are not almonds create an ecosystem that cannot sustain pollinating insects. And even if a helpful population of insects existed it wouldn't be big enough to cope with this once-a-year hyperabundance of flowers.

The solution is even more extraordinary than the problem. They bus the insects in. Every year, from all over the United States, vast lorries transport hives buzzing

Opposite Purposeful beauty: Almond Branches in Bloom, *1890, by Vincent van Gogh (1853-90).*

with bees into Central Valley, and once there, the bees get pollinating. It is the largest managed pollination event in the world: 1.4 million hives are brought in. Most producers need two hives for every acre, at a cost of US$200 per hive. The price has gone up in recent years because of the phenomenon of colony collapse disorder. This is a problem that affects the hives of domestic bees; the causes have not been certainly identified and no straightforward cure exists. Hives often leave Central Valley up to one-third depleted, and it is hard to bring numbers back up again.

Almonds are native to Iran and belong to the genus *Prunus*, along with plum, cherry, peach and apricot. This makes intuitive sense if we concentrate on the blossom rather than the fruit. Technically, the fruit the almond tree produces is not a nut: a nut is defined botanically as a seed inside a hard case which doesn't open naturally to release the seed: acorns, beechmast and hazels are nuts, but cashews, Brazil nuts (see Chapter 14), pistachios, peanuts, pine nuts and walnuts are not. In culinary and informal use, all of the above are of course nuts.

But the drupe that comes from almond trees has been greatly appreciated. The plant was spread first around the Mediterranean and then to other Mediterranean-type climates during the Age of Discovery. Almonds have been cultivated for about 5,000 years; they have an advantage for early farmers in that they grow readily from seed, and don't need the technology of grafting. Almonds were introduced to Spain in the twelfth century; their cultivation is described in Ibn al-'Awwam's *Book of Agriculture*.

Varieties of sweet almonds have dominated the bitter ones, not just for the more pleasant taste but because the bitter varieties contain a good deal of cyanide; a feast of fifty nuts could be enough to kill you. Cyanide is a defence put up against predators by many plants, including apples, peaches, apricots, barley and bamboo shoots, but bitter almonds have a high concentration of the stuff. Many a fictional detective has taken one sniff at the corpse in the library, inhaled the scent of bitter almonds and at once diagnosed cyanide poisoning. Almonds are valued for their health benefits: an ounce can give you one eighth of your daily protein needs, and it is claimed that they lower cholesterol and reduce the risk of cancer. They are rich in unsaturated fats, and the liquid obtained from them – almond milk – is reckoned by some to be healthier than dairy, as well as acceptable to the lactose-intolerant.

In Israel the almond is the first tree to come into blossom, doing so even in late winter. For that reason it has a special significance in Judaism. In the Bible, Aaron's rod was miraculous: the Book of Numbers explains: 'Now it came to pass on the next day that Moses went into the tabernacle of witness, and behold, the rod of Aaron, of the house of Levi, had sprouted and put forth buds, had produced blossoms and yielded ripe almonds.' In the Book of Exodus it is announced that the Menorah, that great lamp of Judaism (see also Chapter 82 on olives), was to have cups in the form of almond blossoms.

Marzipan, or marchpane – almond paste with sugar – is associated particularly with Easter, probably because of the significance of almonds in Judaism. Examples include the German Osterbrot, the Italian Colomba di Pasqua and the British Simnel cake, which bears eleven balls of marzipan, one for each of the disciples who remained loyal after the betrayal by Judas. It is also an essential ingredient to stollen, the Swedish prinsesstårta, pithivier and halva.

Almonds remain remarkable for their pollination event, but they are not alone. Other crops that require pollination if they are to produce food include okra, kiwi, potato, onion, cashew, celery, star fruit, beets of all kinds including sugar beet, mustard, rape, broccoli, cabbage and all the other brassicas, turnips, peas, all kinds of beans, chilli, papaya, caraway, melon, oranges, lemons and all citrus fruits, coconuts, coffee, tobacco, cucumber, squash, carrots, strawberries, cotton, sunflower, flax, lychee, apple, mango, alfalfa, avocado, apricot, cherry, pomegranate, pear, raspberry, aubergine, clover, blueberry, tomato and grape. That list is by no means exhaustive. A study from Cornell University in the United States concluded: 'Without insect pollinators roughly a third of the world's crops would flower only to fade and then lie barren.'

There have been advances in the development of a self-pollinating almond. The Tuono variety can do without bees but it yields an inferior crop. This leaves the almond industry with the classic dilemma of twenty-first-century agriculture. Is the answer to harden the system even more, making it still more remote from the systems of the natural world? Or should the system be softened by a less abrasive, less monocultural post-modern system, one that allows native plants alongside the almond to bring in native pollinators? Opinions remain utterly divided.

EIGHTY-SIX
HEMLOCK

My heart aches, and a drowsy numbness pains
My sense, as though of hemlock I had drunk.

John Keats, 'Ode to a Nightingale'

The chapter is entitled 'Slay It With Flowers'. In its course James Bond, in the novel *You Only Live Twice*, learns that his next enemy is the kingdom of plants. Tiger Tanaka calls it 'the garden of death' and explains: 'This Dr Shatterhand has filled this famous park uniquely with poisonous vegetation.' Then follows a daring break from the narrative, in which Bond reads up about the enemy: a list of twenty-one plant species plus one of fungi, with their nature and toxic properties explained with scholarly detachment over the course of half a dozen pages. It makes for riveting reading: had the reader ever imagined there was so much evil in the world? Only one man can save us: Tanaka tells Bond: 'You are to enter this castle of death and slay the dragon within.'

Many plants contain poisons; many of them can kill humans, though that is not their function. They were not created by the devil, despite many folk names that suggest such things – hemlock has been called devil's bread and devil's porridge – but were evolved as protection from predators. The poisons in the common plant hemlock are there to put off grazing mammals, but they are reliably lethal to humans in comparatively small quantities: half a dozen leaves are usually enough to do the job.

Hemlock is a member of the carrot family *Apiaceae*, which adds a homely touch to the exotic notion of death by poisoning. All parts of the plant are toxic, especially the seeds and the roots. The plant is native to Europe and North Africa, and has been introduced – either accidentally as a weed or deliberately as a garden plant – to Australia, West Asia, New Zealand and North and South America; it can be found in twelve states of the USA. The plant is a biennial and often grows along roadsides. They do well on poorly drained soil near streams and ditches. They are poisonous to all mammals and many other organisms. They have been known to affect cattle, pigs, sheep, goats, donkeys, rabbits and horses. It is particularly

Opposite Deadly: Hemlock, 1854, coloured by Albert Henry Page, after Grandville (real name Jean Ignace Isidore Gerard).

important to make sure there is no hemlock in hay, the dried vegetation that is kept for winter fodder. The plant is more palatable when dried, and more likely to be consumed in quantity.

Hemlock plants grow as attractive white umbellifers – an innocent appearance which adds to their sinister reputation – and can be confused with the humdrum cow parsley. They normally grow up to waist height, but can reach higher. The stalk is hairless, and is green with purple spots: that can be considered a warning sign. The active substance is coniine; the plant's scientific name, *Conium maculatum*, comes from the Greek name for the plant *koneion*, which in turns comes from the word *konas*, which means vertigo or dizziness, the first symptom of hemlock poisoning. It begins with a state like drunkenness, but the fun stops pretty early; the coniine acts on the central nervous system and will cause respiratory collapse within seventy-two hours. There is no antidote of the kind beloved of the authors of thrillers; all that can save a victim is a sustained period on a ventilator. Hemlock is pretty relentless: inhalation of substances released by the plant can be harmful, and skin contact results in irritation. Hemlock has been admired, feared, hated and, naturally, mythologised.

The principal hemlock story concerns the death of Socrates, and it's essentially true enough. Socrates, regarded as the first great moral philosopher of the Western tradition, was found guilty of impiety and corrupting the minds of Athenian youth. He was given a choice: renounce your teachings or accept death. He took the latter option, was given a choice of the means and opted for hemlock. The story is told by his pupil Plato in *Phaedo*, in a famous passage:

> The man ... laid his hands on him and after a while examined his feet and legs, then pinched his foot hard and asked if he felt it. He said 'No'; then after that, his thighs; and passing upwards in this way he showed us that he was growing cold and rigid. And then again he touched him and said that when it reached his heart, he would be gone. The chill had now reached the region about the groin, and uncovering his face, which had been covered, he said – and these were his last words – 'Crito, we owe a cock to Asclepius. Pay it and do not neglect it.' 'That,' said Crito, 'shall be done; but see if you have anything else to say.' To this question he made no reply, but after a little while he moved; the attendant uncovered him; his eyes were fixed. And Crito when he saw it, closed his mouth and eyes.

Asclepius is the god of healing; the enigmatic last words have been much debated. Plato set up Socrates as an ideal figure, as the Great Man: the extent to which he romanticised his death has also been debated; a person in the terminal stages of hemlock poisoning would have been fighting desperately for breath. Pliny the Elder says that drinking wine after ingesting hemlock will operate as an antidote. He also quotes the view that 'if young girls' breasts are rubbed with hemlock they will always be firm. If it is rubbed on men's testicles at puberty it suppresses sexual desire.'

Poisonous plants exert a deep fascination, but it seems strange that many of them are common both wild and in gardens. The sweet berries of deadly nightshade have caused accidental deaths: deliberate ones as well, including, it's been claimed, that of the Roman Emperor Augustus by his wife Livia. The castor bean (source of the emetic castor oil) is a poisonous plant from Africa; it was used to kill the Bulgarian dissident writer Georgi Markov in 1978. The rosary pea produces attractive berries, red with a black dot, and they have often been used as bracelets, necklaces and rosaries. They are perfectly safe so long as they are not scratched or broken, when they can be lethal; jewellers have died while making rosaries from damaged berries.

Popular garden plants that contain dangerous poisons include hydrangea, oleander, daffodil, foxglove, ivy, lily of the valley and rhododendrons: we can all grow a garden of death, and many people do, without the evil intentions of Shatterhand – who of course turns out to be Bond's great enemy Ernst Stavro Blofeld.

The writers of detective stories have always loved a good poisonous plant, and chief of these was Agatha Christie, who was also a great gardener; she had a 30-acre (12 hectare) garden in Devon and established a nursery business. In her vast oeuvre of death and the unmasking of murderers, characters die from yew, foxglove, monkshood, yellow jasmine and of course hemlock. In the Poirot story *Five Little Pigs*, the victim dies from hemlock in a bottle of beer.

EIGHTY-SEVEN
CABBAGE

The time has come, the Walrus said,
to talk of many things,
Of shoes and ships and sealing-wax,
of cabbages and kings.

Lewis Carroll, 'The Walrus and the Carpenter'

At the heart of Freud's system of beliefs is the idea that the traumas you suffer in childhood profoundly affect your adult life. He was more interested in sex than nutrition, but I can vouch for his conclusions if we extend his remit to cabbage. It goes back to my primary school, where the headmistress Mrs Milford would, on occasional whim, attend school dinners and select a pupil who was having difficulty with the food on offer. She would sit over that pupil until the food was devoured. There were always aromatic shredded cabbage leaves, furiously boiled and served in the form of a dome. I had many lunchtimes in Mrs Milford's company, forcing shards of leaves past the gag reflex, with the result that I am unable to consume them as an adult. Perhaps it would have helped had she explained that cabbages are a brilliant and vivid clue to the mechanism and meaning of life, one enthusiastically endorsed by Charles Darwin: 'Everyone knows how greatly the different kinds of cabbage vary in appearance.'

The ancestral cabbage plant is *Brassica oleracea* and it looks nothing like a cabbage: a leggy plant with flabby leaves and yellow flowers. From this one species we have bred red cabbage, green cabbage, white cabbage, Brussels sprouts, broccoli, cauliflower, kale, savoy cabbage, collard greens, kohlrabi and gai lan, or Chinese broccoli. The botanist from Mars who came down to classify these plants would hesitate to put them in the same genus, let alone the same species; the Martian zoologist would have the same trouble with dogs, and for the same reason. But all dogs and all those cabbage-derived plants are the same species: and that's central to any understanding of the way that life operates.

The ancestral cabbage plant grows wild on the coast of south and west Europe. It is a tall biennial with a high tolerance of salt and lime. It gets through the winter by storing reserves of food and that's what makes it attractive as a food for others. It doesn't look like a cabbage or a sprout, any more than a wolf looks like a Yorkshire terrier, but both pairs share the same close relationship. The plant's

Noble cabbage: The Cabbage Field, *1914, by Charles Courtney Curran (1861-1942).*

leaves are edible, with a memorable taste, no doubt attractive enough to those who have been spared the hideous traumas of my childhood. It was first domesticated for certain 3,000 years ago, probably a good deal earlier.

Cabbage was known in Ancient Greece and Egypt. Among its other virtues it was considered the ideal foundation for a night of drinking; a good meal of cabbage, it was believed, allowed you to drink more. Diogenes, the philosopher, cynic and contrarian, is sometimes said to have lived entirely on cabbage and water, teaching by example rather than precept. Cabbage was said to relieve gout and headache.

The ancient Romans took to cabbage in a big way, and praised it as a vegetable above all others. Pliny the Elder begins his discussion of the plant: 'It would be a lengthy task to list the good points of the cabbage.' He lists seven varieties, so the plant was already beginning to show considerable variation under domestication. He mentions a treatise on the cultivation of cabbages by the Greek physician Chrysippus of Cnidus, and reports that cabbages and vines should never be planted together; if you do so, he said, you will get cabbage-flavoured wine, a singularly unattractive prospect.

The round-headed cabbage – the cabbage we think of as cabbage – was first recorded in England in the fourteenth century. It spread east and west through the usual routes, and became important on sea voyages because it's high in vitamin C and therefore a great preventer of scurvy.

Cabbage can be eaten raw, boiled, steamed and roasted. It can also be pickled and fermented, in the forms mostly widely known as the German sauerkraut and the Korean kimchi. Cabbage preserved in such forms has been used to dress wounds and prevent gangrene; these days it is a prized health food, and it is claimed that it is not only nutritious but helps digestion and weight loss, reduces stress, is good for the brain, reduces the risk of cancer, improves the health of the heart and makes your bones stronger. The record size for a single cabbage is 138 pounds 4 ounces (62.7kg).

Cabbages were interesting to Darwin because of their variation. The first chapter of *On the Origin of Species* is called 'Variation Under Domestication' and concerns the way that selective breeding by humans has altered the plants and animals we live with: 'If man can by patience select variations useful to him... should not variations useful to nature's living products often arise, and be preserved and selected?'

He reinforced this point in his book of 1868, *The Variation of Plants and Animals Under Domestication*. This can be regarded as a 1,000-page appendix to that chapter. Darwin was never content with his breakthrough: he had to demonstrate its truth again and again.

'It is an error to speak of man "tampering with nature" and causing variability. If organic beings had not possessed an inherent tendency to vary, man could have done nothing.' The cabbage provides a beautiful example of this. The cultivated forms are not only radically different to the ancestral wild plant; they are different to each other, and in many different ways. If humans can cause such radical changes in a comparatively short time, what might the powers of unaided nature do over a period of geological time?

'Man therefore may be said to have been trying an experiment on a gigantic scale; and it is an experiment which nature during the long lapse of time has incessantly tried. Hence it follows that the principles of domestication are important for us. The main result is that organic beings thus treated have varied largely, and the variations have been inherited.'

I remember being taken to the London Planetarium as a child – a welcome day off from Mrs Milford – and being much struck by the grandiloquent commentary that came out of the darkness. It was all about humanity's eternal quest: we must look further and know more: 'And that,' the voice thundered, 'is the difference between man and the cabbage.'

But if he had read Darwin, he would have had to accept that, essentially, there isn't one. A difference, I mean. We are both the products of selection: we have both survived and prospered because of our innate tendency to variability. Cabbages and kings: what's the difference?

EIGHTY-EIGHT
CASSAVA

*Everyone knew he could foretell war and famine, though that
was not so hard, for there was always a war, and generally a
famine somewhere.*

Mark Twain, *The Mysterious Stranger*

Cassava is the food for the most desperate times: it's been called the drought, war and famine crop. It has kept millions of people alive in the most forbidding circumstances: and yet it is known in Britain as the food of nightmare, at least to all those of a certain generation. In its form of tapioca it was made into a milk pudding, with eggs and plenty of sugar. I have never heard of it being served outside schools, of the boarding or the day variety. In a poll organised by *Good Food* magazine, tapioca pudding was voted the worst school food of all, beating cabbage (see previous chapter) into a second place. It is – or was – both bulbous and slimy, most often known as frogspawn or fisheyes, or, more expressively still, eyeball pudding. I didn't mind it too much myself, perhaps because Mrs Milford saved her energies for the main course, but among my fellow pupils, it provoked the liveliest horror. The fact that many people owed their lives to the plant would not have helped them to get it down.

Cassava is also known as manioc, mandioca and yuca (but not yucca, which is unrelated). In the tropics it is the third most important staple after rice and maize; for 800 million people, mostly in Africa, it is the primary food. It is a woody perennial shrub native to South America, and a member of the spurge family *Euphorbiaceae*. It has long tuberous roots, and it has been cultivated for their sake for about 10,000 years, making it one of the earliest of all crops. It was a staple of pre-Columbian civilisations, but when the Spanish and Portuguese arrived they didn't take to it; it's one of the examples of the Columbian exchange that never made it to the mainstream of European cuisine – outside British schools, anyway. For all that, its uses were pretty clear: it was a readily available source of carbohydrates, offering the biggest yield of carbs after sugarcane and sugar beet.

The Spanish grew it on Cuba, because it was hard to grow wheat there. It was fed to sailors, who didn't like it much, and was taken across the world by European traders visiting Africa and Asia. It has obvious uses as animal feed. It has many remarkable virtues: it is drought-tolerant and grows well on marginal soils. Perhaps

Famine fighter: image of cassava, 1789, by Felix Delahaye (1767-1829).

the most important fact of all is that it has an exceptionally wide harvesting window: you can keep it in the ground for a long time, and the roots, the part you eat, just keep on growing. It has comparatively low labour costs even though it is harvested by hand. That has made it the most marvellous reserve crop: almost literally a hedge against famine. When other crops fail, in times of extreme weather, cassava will keep on loyally growing and will provide the food needed to get through. Cassava has done that time and again across history.

It was an important crop in government programmes designed to deal with the major disruptions to food supply caused by the droughts of the 1980s, and by the weather fluctuations caused by El Niño in the 1990s. It has also been a life-giving food in times of conflict, perhaps most notably in Angola, in the troubles that afflicted the country between 1961 and 1997, and also in periods of strife in Mozambique between 1977 and 1992, and during the Biafran War of the late 1960s. In more recent times cassava has become at least as much a cash crop as a subsistence crop: sold by farmers to feed the ever-growing populations of the cities.

There is a drawback to this: cassava is full of cyanide. As always, it's impossible to imagine how such a dangerous crop entered the human diet: how people discovered that something potentially lethal was also one of the most sustaining food items on the planet. The traditional method used to make the food palatable in Africa is to soak the roots for three days and induce a fermentation. You can also grate the cassava into coarse flour, make a thick paste and then let it stand for five hours. It remains a plant that requires considerable care: in the 2010s there was a case in Venezuela in which incorrectly prepared cassava killed dozens of people. The cyanide has also been employed deliberately, and applied to darts and arrows. This gives the picturesque idea of a plant that enables you to feed your family and kill your enemies from the same bush, but the greater use for such weapons has always been in hunting for food.

You can – I suppose I should add here 'of course' – make an alcoholic drink from cassava; it also makes a good starch for use with clothes. In recent years cassava has been increasingly used to make ethanol biofuels, especially in China. (We will look at biofuels more closely in Chapter 95 on rape.)

The cultivation of cassava brings about problems of habitat destruction – forests are cleared to grow the stuff – and the long-term degradation of the soil. The great but related problem is that the population of Africa is likely to double by 2050. The way that humanity responds to the twin challenges of impoverished soils and rising populations will profoundly affect the viability of the planet as a life-sustaining organism. This brings us to the idea of 'sustainable intensification'. This has always seemed to be an oxymoron, but if we are to deal with the needs of a population that continues to expand, we need to make that concept an inevitable twinning, if not a tautology.

EIGHTY-NINE
ORDEAL TREE

O, my offence is rank, it smells to heaven…

William Shakespeare, *Hamlet*

Perhaps we never really grow out of the binary view of life. If it's alive it's either a plant or an animal, and it's either good or bad. Wheat and roses and oaks are good, bindweed and opium poppies and kudzu are bad. People, nations and religions fall into either of the above categories. We accept such things when we are babes in arms, and though we spend the rest of our lives learning about nuances, I'm not sure we ever take them to heart.

It's easier to think of hemlock and toadstools as unequivocally bad things, almost as mistakes in nature, or as proof of nature's malignity. When we think about it we know that life isn't like that, but not thinking is far more convenient. The fact remains that the actual poison from many poisonous plants have been important to humans: keeping individuals and families and communities fed. It might be claiming too much to say that humanity may not have survived without access to poison, but poison has certainly played its part in keeping humanity going, from hunter-gatherer days before the invention of agriculture, right up to the present day.

The ordeal tree is a classic example of a useful poison. It is named, feared and mythologised for the part it has played in ancient trials by ordeal, and we shall look at these in a moment. But its importance to humanity has not been in the dramatic quest for justice, but for a much more humdrum and homely reason. Like a fish supper.

Hunting for fish with poison has been practised in many societies, important to hunter-gatherers, and later as a useful addition of protein once humans had become an agricultural species. The practice has been pursued in different places with different poisons extracted from different plants. In Africa, the tree of choice for this risky but highly effective form of fishing has always been ordeal trees, trees of the genus *Erythrophleum*; there are species across most of sub-Saharan Africa and more in Asia and Australia. The ordeal tree creates these poisons for the usual reason; for defence against predation, and the poison is found in the highest quantities in the bark and in the seeds.

The basic principle is simple enough: get upstream of a good-looking shoal, sprinkle your poison and then pick up your catch; often enough you can use your

 Pub. by S. Curtis Walworth, Feb.1. 1830. Swan sc.

1

hands for the job. The process is vividly described in the James Bond short story 'The Hildebrand Rarity'; Ian Fleming was a closet nature-writer. 'And then with stupefying suddenness everyone went mad. It was as if they had all been seized with St Vitus dance. Several fish looped the loop crazily and then fell like heavy leaves to the sand.' The bravura description continues for another sixteen lines as a community is wiped out before Bond's horrified eyes.

But there's little room for the softer emotions when survival of self and family are involved. The bark of the ordeal trees and its leaves are gathered and partly crushed in water, and then (these days) placed in a mealie-meal sack, one used for storing ground maize. The most effective way of exploiting the poison is to build a temporary dam; if the river is too wide to make this feasible then a judicious obstruction at a suitable point in the river will corral most of the poisoned fish and allow you to gather them easily enough.

It is, however, a business fraught with danger. It damages the ecosystem, affecting fish beyond your target area and affecting the future availability of fish to the human community; what's more, the stuff is poisonous to humans as well as fish. When it has been used too lavishly or in too concentrated a form, the fish that you catch can hold higher levels of poison than you would wish, enough to cause illness, potentially serious in small children.

That has given rise to a jocoserious piece of advice that's bandied about in parts of Africa: if you come across a dead fish that might be the victim of poisoning, you can make it safe by boiling it in a pot with a stone. When the stone is soft you throw away the fish and eat the stone; the joke continues with the idea that the stone will have taken on delicious fish flavours.

The ordeal tree provides good timber, but that too is not without its risks: the sawdust is likely to irritate the mucous membranes when inhaled. All the same it's good wood for all kinds of joinery, for flooring, bridges, boats and wheel-hubs; it has also been used for railway sleepers (ties). The tree has been grown for this reason and quite often as an ornamental tree, sometimes planted in avenues, because it is a handsome thing; the sinister reputation no doubt adding to its attraction.

Trial by ordeal has been found in many if not all parts of the world, Europe included. It operates on the idea that the guilty can be separated from the innocent by means of a test: reaching into a boiling pot to collect a stone, for example, or taking poison: innocent people will vomit up the poison but the guilty will die. It's not without logic: a guilty person, knowing his own guilt, would be unlikely to accept the offered ordeal; while the innocent person would do so readily. Those who control the dose of the poison to be ingested have the opportunity to take this into account.

Opposite Truth-finder: ordeal tree illustration by Lady Frances Cole, 1830.

My friend Manny Mvula, when working as camp manager in the Luangwa Valley in Zambia, had some money stolen. All the camp staff denied responsibility. He was advised to visit a sorcerer and was given a concoction of poison to be mixed with water. He took this back to camp and showed it to the staff. 'They all but one agreed to take the poison. That one refused because he was the culprit. He later confessed and brought back the money. Not sure what would have happened if he'd drunk it...'

The ordeal tree contains powerful substances and they have been used widely in traditional medicines, so much so that with rising populations ordeal trees are increasingly scarce. The substances obtained from the trees have been used as an emetic, an anaesthetic, against malaria, for pain relief, as disinfectant, to treat dermatitis, inflammation, headaches, gangrene and rheumatism. A large dose in warm-blooded animals (birds and mammals) leads to shortage of breath and seizure; cardiac arrest follows within minutes. But with its judicious use you can catch an armful of fish and your family can survive another week.

NINETY

CHRYSANTHEMUM

A white chrysanthemum
however intently I gaze
not a speck of dust

Basho

The chrysanthemum has a mixed reputation. In a number of European countries – France, Belgium, Italy, Spain, Poland, Hungary, Croatia – it's mostly for funerals. In England it's ever-so-slightly absurd: altogether too florid a flower. In East Asia it's a flower that strikes awe. All over the world it's also a plant of power and significance. It is the source of potent insecticides, telling the story of rising populations over the past century and summing up one of the great dilemmas of the century to come.

My mother hated the idea that she might die in autumn, and that her funeral would take place in a church full of chrysanthemums. She disliked their appearance, the fleshy feel of the flowers and above all the smell, which seemed to her the scent of decay. D. H. Lawrence wrote a classic short story entitled 'Odour of Chrysanthemums', a tale of disillusionment, death and a spilled vase of chrysanthemums. To be fair, others find their scent pleasing, earthy and herby.

In the P. G. Wodehouse novel *Leave It to Psmith*, the eponymous hero goes to a secret encounter wearing for recognition a pink chrysanthemum in his button hole. Alas, the person who arranged the meeting had been thinking of carnations.

'I AM wearing a pink chrysanthemum. I should have imagined that was a fact the most casual could hardly have overlooked.'

'That thing? … I thought it was a kind of cabbage.'

The chrysanthemum is native to East Asia, where it is held in higher regard… but now we come to one of those problems of names that please the precise minds of taxonomists while spreading confusion through the rest of us. The genus *Chrysanthemum* belongs to the family of daisies, or *Asteraceae*, which contains more than 1,500 genera. The florists' chrysanthemum used to belong, rationally enough, to the genus *Chrysanthemum*, then it was kicked out, and assigned another genus. Then it was put back in again, at least by some, but classification is still a contentious issue, serious enough if you are interested in the way life operates. The genus *Chrysanthemum* includes, it's usually agreed, species that not only have big

LUCIEN DAVIS·R·I·

showy flowers but also powerful insecticidal properties. This includes the species *Chrysanthemum pyrethrum*, which is referred to informally, along with related species with a similar property, as pyrethrum. If you've ever been in a room with a burning mosquito coil you have felt the benefit of pyrethrum.

The flowering chrysanthemum is so important in Japan that the institution of the Japanese monarchy is known as the Chrysanthemum Throne. There's a chrysanthemum on the country's imperial seal and a national celebration of Chrysanthemum Day, the ninth day of the ninth month, when chrysanthemums are at their best; they are, as my mother was well aware, plants that bloom late in the year.

Chrysanthemums were cultivated in China as early as 1500 BC. They were regarded as one of the Four Gentlemen of Chinese art, the others being plum blossom, bamboo and orchid. They reached Japan by the eighth century and soared in popularity during the Edo period, that is to say, most of the seventeenth, eighteenth and nineteenth centuries. It's been reckoned that there were at least 500 cultivars by 1630, and today by some accounts there are as many as 20,000. The plant is a hardy perennial that survives the northern winter. The name, much misspelt in the days before spell-checkers, is from the Greek and means 'golden flower'.

The chrysanthemum species known as pyrethrum might be worth gold to many people. Certainly they have changed lives. Their insecticidal qualities have helped to reduce cases of malaria by keeping mosquitoes away, and they have allowed farmers to produce more crops on the same land by killing insects that damage them. Insecticide can be made from pyrethrum flowers simply by drying and crushing them and then mixing the dust with water. Pyrethrum can be used as companion planting, their presence protecting food-plants from insect damage. The use of pyrethrum is permitted in certification for what is called organic farming.

The plant's active substance, pyrethrin, has been synthesised to create pyrethroids. These have a higher level of toxicity: they are better at killing insects and easier and cheaper to produce. Synthetic insecticides are now readily available almost everywhere, and it's no exaggeration to say that they have changed the world. They started from an original idea from the plants, plants like pyrethrum. They come in three types: those that must be ingested and damage the digestion, those that must be inhaled and damage the respiratory system, and those that operate on contact and damage the outer covering, the exoskeleton. Pyrethrum is in the last of these categories. Most synthetics work, at least to an extent, in all three ways at once. They also last longer between applications; natural insecticides tend to have short effective life.

Synthetic insecticides came into widespread use in the mid-twentieth century. Bill Bryson's book *The Life and Times of the Thunderbolt Kid* celebrated 1950s

Opposite Heady flowers: In the Conservatory, c.1920, by Lucien Davis (1860-1941).

America, in which, as the blurb for the book states, 'everything was good for you, including DDT, cigarettes and nuclear fallout'.

Synthetics were a crucial part of the great leap forward in agriculture. It's been estimated that in some places, fields had a 50 per cent increase in yield between 1945 and 1965 because of the newly available insecticides. More food meant that we could have more people on the planet, especially as insecticides also slowed down transmission rates of malaria, yellow fever and typhus. Insecticides enhanced the survival rate of humankind: populations have risen as the inevitable result.

The downside to the proliferation of insecticide is well-known enough: environmental contamination, the development of resistance among the target species and the inadvertent death of non-target species – some of which, like pollinating insects, are essential to many good crops, as we have already seen (Chapter 85 on almonds). Until recent times the loss of nature and natural diversity had been looked on as collateral damage: regretted only by a few people who could be written off as sentimentalists. There is now a fear that monocultural intensity, created to feed an ever-increasing population, will irretrievably compromise the planet's future as a food-producing organism. Once again we are facing a terrible dilemma.

We have operated for years on the principle that human ingenuity will find a solution, because it always has done before. Now we must ask if this approach is still viable, with more and more human mouths to fill. Do we bet on nature, or on humanity's ability to take nature out of the equation?

NINETY-ONE
MYCORRHIZA

I talk to the trees…

Alan Jay Lerner and Frederick Loewe, sung by Clint
Eastwood in the 1969 film *Paint Your Wagon*

It's possible that your eyes have glazed over at this chapter title, so rich in late-alphabet letters. Perhaps you think that I am running short of options for these late chapters and getting desperate. If so, think again: I am, of course, saving the best till last. There's an argument for saying that this is the most important chapter of the entire book. It's also one that fills the mind with wonder, just as it has filled the world with life.

Mycorrhizal fungi live in and round the roots of plants. It's not a species. It's not any kind taxonomical category: it's an ecological description, so it tells us about a way of making a living. Mycorrhizal fungi are associated with about 90 per cent of all plant species, and both parties benefit from the association. The fungi get food, because, like us animals, they are consumers; only plants are creators. The plants benefit from the water the fungi bring, and also and especially from the nutrients like nitrogen and phosphorous, which the fungi make available to the plants, and would otherwise be inaccessible. In a forest they create a system that allows the trees to communicate, sending messages, alarms, assistance – and on occasions hostile substances, in a manner that has fancifully been compared to cybercrime.

To understand this we must once again recalibrate our notion of what constitutes a fungus. The mushrooms in the risotto, the menacing toadstool on the forest floor: these are the fruiting bodies of fungi, the stuff that pushes out the spores and allows them to become ancestors. To confuse them with the entire living organism is like mistaking an acorn for an oak tree. The heart and soul of a fungus in is the threads called mycelia.

The first thing to grasp about mycelia is that it comes in quantities that are almost beyond calculation. Here's one estimate: in every teaspoon of soil there's a kilometre (more than 1,000 yards) of mycelia. Threads of mycelium are often so fine they are invisible to the naked eye. Here is life beyond our easy understanding, but the way it operates is essential to us all. It's perhaps a little galling to find that you depend on something that you don't know about, can't see and can't easily understand: but life has always been like that.

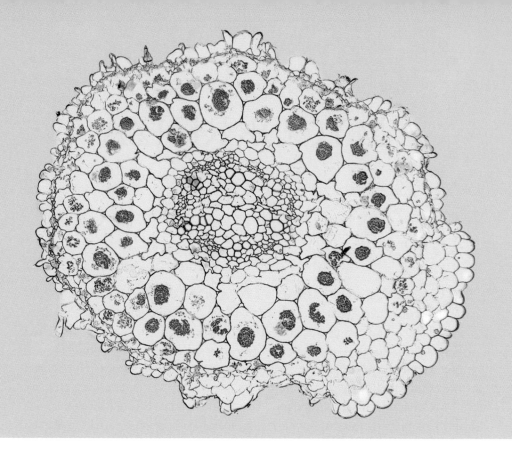

Wood wide web: light micrograph of a cross-section through a root containing fungal hyphae.

The network of mycelia enriches the lives of both consumer and consumed, and it goes a long way to make the existence of both possible. It's been suggested that the invasion of the land by plants 450 million years ago was made possible by the existence of fungi. The fungal threads stretch out under the ground in all directions. In a forest they link one tree with another: like colleagues in Lockdown the trees cannot communicate with one another directly, but must use an external medium to help do the job. The fungal network acts like a Zoom meeting and has been called, charmingly, the Wood Wide Web.

The idea of trees talking to each other sounds alarmingly like New Agery: but the unexpected truths that have been revealed are hard-edged. For example, some big woodland trees will transfer their nutrients, via the fungi, to nearby plants in sub-optimal conditions: aiding, for example, a tree with less access to life-giving sunlight than the giant. It's possible that by this means a parent tree can succour its own progeny. It's also true that a wood benefits from its own extensiveness: the bigger the wood, the safer it is for every tree during episodes of extreme weather. It's in every tree's own interests to encourage other trees.

When a tree gets a disease, the information about its condition can be carried to other trees by means of the fungal communication system, and that allows them to boost their immune systems in preparation, producing defensive chemicals: pre-emptive medication that can be compared to vaccination. Within a short space of time every tree nearby is better able to cope with the disease.

The same principle works when it comes to dealing with hostile animal species. It's been demonstrated in laboratory conditions that broad bean plants linked by fungi inform each other when one of their number has been attacked by aphids. These neighbouring plants then produce chemicals that protect themselves against the aphids: should they arrive, the plant is ready. They don't have the stuff already because that would be uneconomic: a waste of energy, since there might not be an aphid attack.

But not all this communication is benign. Plants can also send toxins through the fungal network: they can do so to a competing neighbour, making it hard if not impossible for the plant to establish itself. Perhaps everywhere you find co-operation you also find rivalry: many plants benefit from each other's proximity, but are also in competition for the same resources. Co-operation and rivalry are sometimes inextricable.

The fact that most plants do better with a mutualistic relationship with fungi has obvious implications for agriculture. A good fungal network will bring nutrients to a crop without the same need for chemical fertilisers, while at the same time providing protection from soil-borne predators. Since their invention we have traditionally preferred the application of agrochemicals, which are efficient and comparatively cheap. Their production is tied into the production of greenhouse gases, which is troubling enough, but they also tend to leave the soil worse off than before, once the nutrients have gone. This inevitable train of events eventually requires a terrible decision to be made. Do we now add fertilisers in greater quantity? And do we now expand our growing area, taking down the surrounding trees?

These frightening questions can be answered by the introduction of mycorrhizal fungi, which has a great ring of good sense about it. But its use is both time-consuming and labour-intensive: it is easier, much easier, to go for the chemicals and get the harvest in for that year at least.

We need vegetation. We need vegetation across the world, not just to eat but to breathe. Plants that fill our atmosphere with oxygen: it makes sense to keep our breathing apparatus in good repair. The existence of mycorrhizal fungi makes it possible for plants to thrive, and therefore, at one step distance, we owe our existence to these little white threads. Their existence and their purpose have unleashed new ways of understanding the way that the planet operates. That gives us a better idea of how best to look after the place. It also excites that sense of wonder and delight: a joy in finding that the world is more complex than we are capable of understanding. Perhaps we will find that the key to survival lies in that sense of wonder.

NINETY-TWO
EUCALYPTUS

Once a jolly swagman camped by a billabong
Under the shade of a coolabah tree…

'Waltzing Matilda' by Banjo Paterson

The Greek word for home is *oikos*. That gives us the two words that define the public life of humans, and they have been shaping the planet for the past 12,000 years, ever since the invention of agriculture. These words are economy and ecology. From the closeness of their origin you might expect them to be almost the same discipline: sister sciences, two sides of the same coin. Perhaps that's what they should be, but as we have seen again and again in the course of this book, they are frequently used to express values and worldviews in total opposition. That is made very clear by eucalyptus.

There are getting on for 700 species in the genus *Eucalyptus*, and they go a long way towards defining a nation. Almost all of them are native to Australia and nowhere else. Just fifteen species grow naturally in the islands to the north, and of these, six are also found in Australia. The northern-hemisphere visitor to Australia encounters a strangely disconcerting visual impression on looking out across the Australian landscape, even in city parks: it looks pretty familiar, but something seems not quite right. That's because most of the trees are eucalyptus, and they look subtly different to the trees people from elsewhere are used to.

The early European settlers who tried to paint this landscape struggled with it. In the Art Gallery of New South Wales in Sydney the Australian landscape is there all right, but the trees look distinctly English. Eucalyptus leaves tend to hang downwards, giving trees a different shape and, for that matter, a mostly indifferent shade; these eighteenth- and early-nineteenth-century painters struggled with them.

The genus *Eucalyptus* includes small bushes, mallees – woody plants that are multi-stemmed from ground level – and trees, including the tallest trees in the world. These giants belong to a species, named by English settlers with a mixture of nostalgia and indifference to the science of taxonomy, as the Australian mountain ash (or *E. regnans*); the record is 330 feet (100.5 metres). Half a dozen other species routinely top 260 feet (80 metres). Eucalyptus plants tend to produce a powerful oil in their leaves, and when the bark is damaged, they emit a sticky

Outback: The Settler's Home, *1931, Kiewa Valley, Victoria, Australia, by Harold Herbert (1891-1945).*

self-healing substance called kino, or less formally gum, which is why eucalyptus trees are called gum-trees; the coolabah tree, brief shelter of the hapless swagman, is just such a tree.

The plants are eaten by koalas and possums and they fuel many insects with nectar-rich flowers. They have a striking property to eyes unused to them: their bark is always coming off. They grow a new covering every year, and in many species the old bark is discarded in flakes and strips.

They are adapted to a fire ecology, many of them regenerating from buds beneath the bark or from fire-resistant seeds. In the natural cycle, fire is an opportunity as well as a disaster: swift regeneration is followed by swift growth. Many eucalyptus species burn very readily, and that gives Australian bushfires their peculiar virulence. The trees seem almost to revel in fire: their powerful oils discourage fungi, and so they carry a good deal of dead dry wood, rather than having it digested by fungi. The oils they contain add to their flammability; in the oil-rich air of the forest canopy fires leap readily from crown to crown. In the bushfires of 1974-75, 15 per cent of Australia was burned – 290 million acres (117 million hectares) – though almost all in places far from human habitation. The fires of 2019-20 destroyed 5,900 buildings and affected more than 3 billion terrestrial vertebrates, mostly reptiles, and probably pushed some species into extinction. Fires are getting more frequent with changes in climate.

Australia was built from eucalyptus: houses, furniture and boats. The indigenous peoples used eucalyptus for fire, tools and weaponry, including their famous boomerang, which comes in two types: the hunting kind, which goes in a straight line, and the returning kind, which is recreational. The wooden tool called the woomera allows a spear to be thrown further and faster by acting as an extension of the human arm; the traditional musical instrument the didgeridoo is also made from eucalyptus.

Eucalyptus has been prized for its health-giving properties, mostly the species E. globulus, grown for its essential oil, which is gained by steam-distilling the leaves. It's been used for cleaning, as a solvent, in toothpastes, as a decongestant and as a mosquito repellent. It's also said to reduce stress, relieve pain and treat dry skin.

Some eucalyptus species make good timber, and they are also useful for wood-pulp. They tend to be very fast-growing, and tolerant of poor conditions; some species are able to use their bark for photosynthesis. These qualities have made eucalyptus the perfect choice for commercial tree plantations, though most species can't deal with frost and so they are restricted to warmer countries. Eucalyptus plantations have been established in many places, including California, Spain, Portugal, Africa, East Asia, South America, South Asia and China. Eucalyptus is an important cash crop in poorer nations; it gratifyingly regenerates from the root when felled. As a result, eucalyptus species are the most widely used plantation trees in the world.

They seem almost perfect trees, rescued from a remote continent to do the bidding of humankind in a changing world. There is, of course, a payback. The tendency of eucalyptus trees to burn is only one of them. They grow deep into the soil to extract and therefore deplete groundwater; they encourage erosion. They flourish at the expense of native species of plants, and that depletes the biodiversity of native animals. They become invasive for the usual reasons: there are no pathogens or consuming animals in the local ecosystem to inhibit them. But that hasn't stopped the planting. It's reckoned that up to 47 million acres (19 million hectares) worldwide are given over to eucalyptus plantations. In many places eucalyptus expresses the victory of economy of over ecology.

However, in the twenty-first century there is a new way of reckoning these two things. It is based on the notion of natural capital: that what has been termed 'ecosystem services' – things provided free by nature – are actually worth a great deal of money, sums that can be quantified. The most obvious example is pollination: if you kill off all your insects you have to spend money to get your crops pollinated: it's been calculated that the work of insect pollinators in the USA in 2012 was worth US$34 billion (see Chapter 85 on almond). The principle holds good right across the world and the entire range of human activities: if you destroy the places of resort and recuperation that humans seek in the natural world, they get more stressed and require expensive treatment and drugs.

The division between ecology and economy is at base a question about time. Economy is about the quick fix: the modern human belief that if you look after the economy everything else will fall into place. Ecology is about generations. Our choices are before us and they define our species and our world.

NINETY-THREE
COCA

You just picked up a hitcher
A prisoner of the white lines on the freeway

'Coyote' by Joni Mitchell

Throughout history humans have tried to make bad things less bad and good things still better by vegetal and fungal means, or both together, with alcohol. Chocolate, heroin, LSD, coffee, tea, cannabis, a packet of Marlboro and a bottle of Chateau Margaux '53 (best of the century): all these change our consciousness and they all come from plants and fungi.

Good medicine: Vin des Incas, *1897, by Alphonse Mucha (1860-1939).*

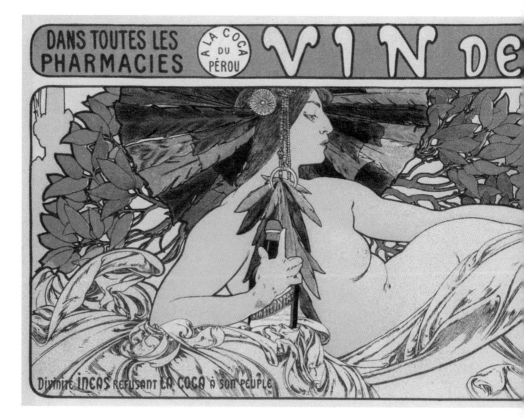

Every society finds this process problematic and complex. There are mind-altering substances that you can buy at supermarkets while the possession of other mind-altering substances will get you sent to prison. Alcohol has probably ruined more lives than any other of these drugs, but it's central to life in many countries, and what's more, the income from taxes on alcohol helps those countries to function. Islamic cultures take a different view, and in some Muslim countries the possession and consumption of alcohol is illegal. It's now accepted globally that tobacco is bad for you, but most governments still profit from it. There is no serious thought of making it illegal. As we have seen in Chapter 12, some countries have changed their mind about cannabis: it used to be evil but now it's good for you – and the state profits. Heroin and LSD are illegal almost everywhere and attempts to enforce the law are highly expensive. Instead of getting money from a tax on heroin, governments are obliged to spend money trying to suppress it.

Nations across the world ban cocaine, which is derived from the leaves of the coca plant. As a result there is a huge illegal trade in cocaine, fuelled by a thousand songs in its praise. Even the songs that point out the drug's drawbacks make it sound pretty sexy. David Bowie, Fleetwood Mac, Eric Clapton, the Rolling Stones, Steely Dan: overt and covert references to the stuff are there for us all, in songs we can hum all day.

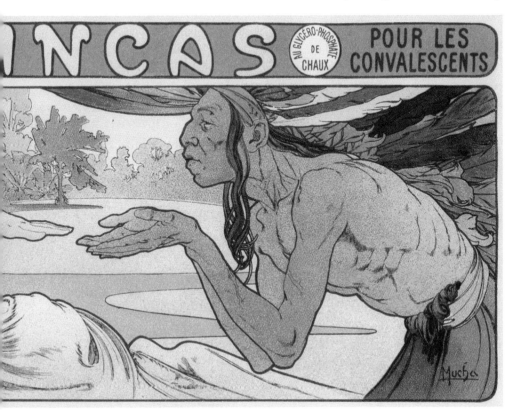

All this helps to drive an illegal market: it's been calculated that 1 per cent of the entire world's GDP is generated by the illegal drug trade. Cocaine begins in the leaves of a plant, and the way they protect themselves against attack by insects.

There are two species of coca plants, and each comes in two cultivated varieties in the family *Erythroxylaceae*. They grow wild in the eastern Andes but have been domesticated for millennia in northeast Argentina, Bolivia, Colombia, Ecuador, Peru and, more recently, Mexico. The plant grows in a bush up to 10 feet (3 metres) in height, producing small clusters of flowers and red berries. The substance that people seek is in the leaves, and for centuries these have been chewed. That process numbs pain and hunger, while giving the chewer a feeling of energy and clarity of thought. In this form, the plant is not addictive physically or psychologically, and in South America it's widely chewed or drunk as tea. Some date the cultivation of the coca plants back 8,000 years; coca leaves were placed in the mouths of the mummies of the distinguished Inca dead. In some cultures coca was restricted to the ruling classes; it was also used for religious purposes.

When the Spanish reached South America in the sixteenth century they objected to the religious use of coca, but saw the point of allowing those in forced labour to chew the stuff, because it made them work harder. The leaves were brought back to Europe but failed to excite much interest at first. In the nineteenth century Dr Paolo Mantegazza changed that by writing in praise of its effect on cognition, and the stuff had a brief vogue. Angelo Mariani marketed a mixture of Bordeaux wine and coca as a tonic. In 1886 an American pharmacist named John Pemberton came up with a drink based on sugar syrup and coca. It was sold at soda fountains as Coca-Cola and was manufactured in this form until 1903.

In 1859 Albert Niemann extracted the active alkaloid from the leaves and gave the world cocaine. It was used an anaesthetic; in 1905 procaine was synthesised by Alfred Einhorn and given the trade name Novocaine. Einhorn hoped it would be used for amputations, but surgeons preferred a general anaesthetic. It was taken up in dentistry and most of us have cause to be grateful to it.

Cocaine crops up in the second of the Sherlock Holmes novels by Arthur Conan Doyle, *The Sign of Four*. The novel begins with Holmes, bored and listless, shooting up cocaine in the sitting room, 'a seven per cent solution'. At the end, with everything settled, Dr Watson says: '"You have done all the work in this business. I get a wife out of it, Jones gets the credit, pray what remains for you?" "For me," said Sherlock Holmes, "there still remains the cocaine bottle."'

Sigmund Freud was a great enthusiast for cocaine, and in 1884 he published *Uber Coca (About Coca)*, which he described as 'a song of praise to this magic substance'. He said that it was a tonic for depression and impotence.

The use of cocaine was forbidden in the United States under the Harrison narcotic act of 1914, and the illicit trade was underway. In 1934, Cole Porter wrote the song 'I Get a Kick Out of You' for the musical *Anything Goes*:

Some get a kick from cocaine.
I'm sure that if
I took even one sniff
That would bore me terrifically, too…

The drug really came into vogue in the 1970s, in the post-hippy era. By this time drug-taking was about pure pleasure, no notion of seeking the ultimate truth. It was especially popular in young businesspeople, the sort that hippies had previously despised as non-drug-taking straights, and was often snorted through a high-denomination banknote. It was a drug about confidence and sexual potency. It was also expensive, a symbol of status. In a rising culture of materialism, it was, it seemed, the perfect drug.

A growing market created ever greater supplies and in the 1980s the price fell by about 80 per cent. It was also sold in a new form. Crack cocaine is created by mixing powdered cocaine with ammonia and water and then boiling it down into solid chunks. It could then be smoked for a briefer but more intense high. Its convenience and high addictive potential made it a perfect product, at least for those on the selling side.

The production of cocaine has been linked with international politics. The CIA have been accused of backing cocaine production in Nicaragua. The illegal trade has created fantastically wealthy men, including Pablo Escobar, who ran the Medellín Cartel of Colombia and was said to be worth US$30 billion at his death in 1993; and Joaquin 'El Chapo' Guzman of the Sinaloa Cartel, who was rated by *Forbes* magazine as one of the four most powerful men in the world between 2009 and 2013.

There have been many international attempts to stop the trade, but it continues, as trades will when there are people who wish to buy and people who wish to sell. Illegal plantations of coca have been sprayed from the air with glyphosate. A new variety of coca plant has been discovered; one unaffected by glyphosate, so spraying is a positive benefit to it, destroying competing plants. It's not clear if this has been developed by careful breeding or by genetic modification. It is known as *supercoca* or *la millionaria*.

NINETY-FOUR
MAIZE

The corn is as high as an elephant's eye…

Oscar Hammerstein, 'Oh, What a Beautiful
Morning', from the musical *Oklahoma!*

Maize is perhaps the most divisive crop in the world. Maize, called corn in North America and Australia and mealie in much of Africa, is grown on every continent except Antarctica. In much of sub-Saharan Africa it's the most important food of them all, and when the maize crop fails, famine follows. In other parts of the world it is an aspect of privilege. In Britain it's often grown for entirely frivolous reasons.

In Zambia the basic daily food is *nshima* and relish: a mouldable porridge of maize meal served with a stew or soup to add flavour and protein. It's usually eaten from a communal dish: a daily ritual of sharing and staying alive. Very little of the maize grown in the United States is eaten by humans: it goes to feed domestic animals and to make fuel. In Britain it's grown to shelter pheasants, so that they can be shot for pleasure. No crop more clearly defines the difference between rich and poor.

The ancestral maize plant grows wild in Mexico; natural stands can reach 42 feet (13 metres) in height. It is an annual grass, like wheat, rye, barley, rice and sugarcane. It was probably first cultivated in the Tehucan Valley, where, unusually, it's likely that it spread from a single ancient domestication event, rather than several roughly simultaneous ones. It was a useful plant all right, but rather a business: each plant produced a single small ear of corn and that only about an inch (2.5 cm) in length. Potato (see Chapter 32) was the staple: maize was a nice bonus. It became part of the important planting strategy known as the Three Sisters, along with beans (see Chapter 60) and squash: beans fix nitrogen, maize offers support for climbing beans and squash offers ground cover and therefore inhibits competing plants.

Maize changed under domestication, and became a great deal more productive. On modern maize plants the cobs can reach 7 inches (18 cm), and grow at a rate of ⅛ inch (3mm) a day, and will usually have two cobs per plant. Maize was part of the Columbian exchange, and was brought to Europe by the Spanish; from there it spread out across the world. The Spanish didn't rate it as food. There were

Food for all: Mexican mural, market with maize on sale, by Desiderio Hernández Xochitiotzin (1922-2007).

two reasons for this: firstly, only bread made from wheat was acceptable for communion, and that was central to life and culture. And second, there was a belief that food made you what you are: if you started to eat the food of the indigenous people, you would become like them.

Maize was an important crop in Mexico and South America and remains so today, supplying carbohydrates in tortillas and tamales. It is also eaten as porridge: this includes the famous or dreaded hominy grits of the American South. You make this by cooking the corn in alkali water. It's called grits for the best of reasons: it's gritty in texture. The dish can be eaten sweet or savoury. As Mike Heron sang in 'Warm Heart Pastry':

Hey I'm a hungry man and you know I ain't talking about grits.

In Italy, maize meal is called polenta, and is used to make a cakey accompaniment to savoury dishes. The Portuguese brought maize to Africa: the fact that the crop is versatile and adaptable, tolerant of many different climates, made it a success. It doesn't cope well with cold and needs moisture: prolonged dry weather will cause the crop to fail and so trigger famine in vulnerable places.

In 1894 John Kellogg and his brother Will Keith were running a Battle Creek Sanatorium in Michigan. They left some cooked wheat out overnight and it spoiled but, being on a tight budget, they fed it to their inmates anyway. But before they did so they forced it through rollers, toasted it and served it with milk: perhaps a little to their surprise, it was at once popular. They had invented breakfast cereals. They tried other raw materials and in 1898 found that maize worked best. Now they had invented Corn Flakes. It is often stated that Kellogg, a Seventh Day Adventist, designed and promoted Corn Flakes because he believed that consumption of bland food reduced the urge to masturbate, but that's usually considered an internet myth. What's not mythology is the sugar content; Corn Flakes contain 8 grams of sugar per 100 gram serving; in Frosties the sugar content is 37 grams.

Maize feeds about 1.2 billion people in Sub-Saharan Africa and Latin America, where it supplies roughly 30 per cent of the average person's calorie intake. In the United States very little maize is eaten by humans, despite the great amount that is grown. Of the resulting crop, 40 per cent is grown for ethanol, to make biofuels, and 36 per cent is fed to animals. It is also used to make bourbon whiskey. An acre of maize is capable of delivering 15 million calories of human food in a year; in the USA such an acre actually delivers 3 million; most of the crop is fed instead to cars and domestic animals.

The animals are then fed to humans, but this not an efficient process. It comes down to what's called a Food Conversion Ratio: the ratio of food consumed by the animal when compared to the amount of food that ends up on plates. Chickens are the most efficient, requiring at best only twice as much food as they actually supply; with cattle it averages out at about ten times. Here's a rough guide to Food Conversion Ratios:

Chickens	2x-5x
Pigs	4x-9x
Cattle	6x-25x

In Britain a good deal of the countryside is managed so that people can shoot pheasants, an exotic species of bird that in natural circumstances would be no nearer to Britain than the Black Sea; annually 50 million pheasants are released in Britain. Stands of maize are grown to provide both protection and food for the birds. Sometimes the shot birds are eaten by humans; many are simply buried.

Modern maize is a cultigen: that is to say, it's incapable of surviving in the wild. The cobs of cultivated maize don't fall off by themselves, so the plant can't propagate itself. There is a variety with a high sugar content, usually called sweet corn. Genetically modified maize has been developed and by 2016, 92 per cent of all the maize gown in the United States and Canada was GM, 33 per cent of the world crop. GM maize is herbicide-tolerant, which means you can use more herbicide on your land to keep down competition and create a true monoculture.

Meanwhile, in Africa, maize is a matter of life and death. Tonight a billion people – the lucky ones – will sit around a pot of maize meal and relish.

NINETY-FIVE
RAPE

'A Transport of Delight'

Michael Flanders, title of a song about a London bus

The invention of agriculture was the most significant event in the history of the planet since the meteorite that struck the earth 65 million years earlier and did away with the dinosaurs. No need to gather it: grow it instead. Why wait on the earth's convenience when you can change the earth? It was a process that began 12,000 years ago, as we have seen many times over in these pages, and it probably did so in many places at roughly the same time. In the course of those 12,000 years – a brief flicker in geological terms – humans have reshaped the planet. We have done so to grow food for ourselves, to grow food for our domestic animals, to enjoy mind-altering properties of plants and fungi, to cure our ills and to produce building materials.

Now things have gone a step further. Humans are beginning to grow plants to power their supercharged existence. We are running out of the fuels that come from fossilised plants and other organisms – coal and oil – and we are now aware of the damage their use has done to the planet. So we are beginning to grow plants and use them as fuel – fuel to power mighty engines – without the inconvenience of waiting millions of years for them to fossilise. We can grow power in our fields.

In Europe, rape is a favourite crop for doing this, just as maize is in North America. The name of the plant is, however, uncomfortable. Rape is a brassica, so related to cabbage; *rapum* is Latin for turnip, hence rape the plant. In Latin, *rapere* means to seize, and the derived English word has come to mean sexual violation. Euphemisms have been created as a result: the plant is sometimes referred to as oilseed, rapeseed, oilseed rape, or colza; in North America the oil produced is called canola.

Rape, like cabbage, is a member of the mustard family and has been cultivated for about 6,000 years, initially in India, from where it spread out first to China and Japan. It was used as a break crop: one that was grown and ploughed back in, between the growing of more useful crops. The process keeps weeds down and improves the soil: rape is a classic example of good husbandry.

The crushed seeds produce oil which is good for lubrication and was used for machinery during the Industrial Revolution. It is not good for human consumption

Yellow landscape: rape fields painted by Eve Masur.

because it contains large amounts of erucic acid, which damages cardiac muscles. But in 1973, rapeseed oil suitable for human consumption was developed in Canada: Canada Oil Low Acid, or canola. With this, the production soared in the developed world; it was backed by generous subsidies in the European Union. The finished product was usually marketed as 'vegetable oil'.

As a result the colour and the air of the European countryside were transformed. A field of rape in full flower is a lavish and startling yellow, and in some places such fields even attract tourists. Japanese visitors make detours to view British rape fields in bloom, a startling change from cherry blossom. The considerable acreage given over to rape cultivation fills the air with a peculiarly sharp pollen, to which many people are allergic. I once had a horse who suffered from rape allergy; in the flowering season she was obliged to wear a mesh mask.

Meanwhile, with the growing crisis of climate change and the diminishing quantities of fossil fuels, new ways of powering the human world were sought. One answer is biofuel. Rape produces more oil per unit of land than its competitors like soy; the by-product of rape-oil production is crushed seeds, which make a high-protein animal food, comparable with soy, which can be fed to cattle, pigs and poultry. Rape also has a lower gel point: it goes solid at a lower temperature than its competitors. This is an advantage because a fuel that has turned solid can no longer be pumped.

Biofuel is not exactly a new idea: we have been burning wood for countless millennia, and wood has been used to power railway locomotives. But the pace hotted up with the development of liquid fuels that drive engines. These come in two main types. The first is ethanol, which is derived from the fermentation of crops like maize and sugarcane; this is then blended with petrol (gasoline). This approach is favoured in the United States and Brazil. In Europe the preferred approach is biodiesel, which is extracted from oily plants like rape, soy and oil palm and mixed with petroleum-based diesel.

The arguments in favour of biofuels are attractive: cleaner fuel with fewer emissions, no more expensive than petrol (gasoline) and every chance it will get still cheaper with greater production; engines that use biofuels last longer because the substance does a better job of lubrication; it is easy to source; above all it is renewable and will be available for as long as we have earth to grow plants in. There are claims that it reduces greenhouse gases by up to 65 per cent. And perhaps most seductive of all, it reduces our dependence on fossil fuels, which in most countries have to be imported.

The counter-arguments are considerable. The first is the high cost of production: converting plants to liquid fuel is a complex business. Biofuel crops are grown in a monoculture, and that, as we have seen many times in this book, is a fraught process, one that requires more and more human input, in terms of fertilisers, herbicides and pesticides, which in turn leads to agricultural pollution of such wild land as we have left. There is the question of whether we should use agricultural land for non-food crops, in a world with an ever-expanding human population and ever-increasing food shortages. The manufacture of biofuels requires fuel in itself, and it also uses a great deal of water; this leads to water pollution and damaging emissions. The pressure on land for fuel as well as for food crops will put still more pressure on wild land, which would be increasingly required and converted for agricultural use of one kind or another.

All of this shows how horribly hard it is to do the right thing. OK, we need to cut down our use of fossil fuel, and now we can. Now we are told that this way of reducing fossil fuel dependence is not the perfect answer. In many ways it's as bad if not worse than where we were before. This is not what we want to hear.

NINETY-SIX

ALGAE

My formula for success is: rise early, work hard, strike oil.

John Paul Getty

Certain questions are better unasked: not because they are indiscreet but because no satisfactory answer exists. Is it art? Are humans apes? Are birds dinosaurs? Are algae plants? Well, it all depends on who you ask. And what kind of mood they are in. Yes, but then again – no. All in all, it is and it isn't, if you see what I mean. And of course, it all depends on what we mean by the word algae. The one thing we can be absolutely certain about is that without algae the world we live in would be utterly different, because without algae there would be no crude oil in the ground.

Algae is not a precise taxonomic term: it's used informally for all kinds of living things that photosynthesise: that is to say, that feed directly on sunlight. Many of them are microscopic and comprise a single ceil for each individual; on the other hand, and depending on who you ask, seaweed is also a form of algae, and fronds of giant kelp can reach 160 feet (50 metres) in length, so you'd be reluctant to class them as bacteria. A rough definition: algae comprises aquatic photosynthesising eukaryotes. Eukaryotes are living things whose cell or cells each contain a nucleus. That includes us, of course. Many of the species that we lump in with algae aren't closely related to each other.

We talked in an earlier chapter about climax vegetation, and the way that a closed-canopy oakwood (Chapter 9) is the climax vegetation of, for example, lowland Britain. You can discover the nature of climax vegetation by filling a jam jar with water and leaving it: after a week or so it will create a green scum. That is algae, and it's the climax vegetation of the environment you have created.

Algae do photosynthesis, which makes you want to put them in with the plants, but they lack many if not most of the attributes of most plants: they lack true roots, stems, leaves and a vascular system, as do mosses and liverworts. Algae can form colonies, and work together for each other's mutual benefit. They can also form important symbiotic relationships with other organisms. Coral reefs are formed by colonising coral species that don't just hunt for passing specks of food; they also form a powerful relationship with photosynthesising algae. In return for protection the algae provide the corals with food. That's why coral reefs are found

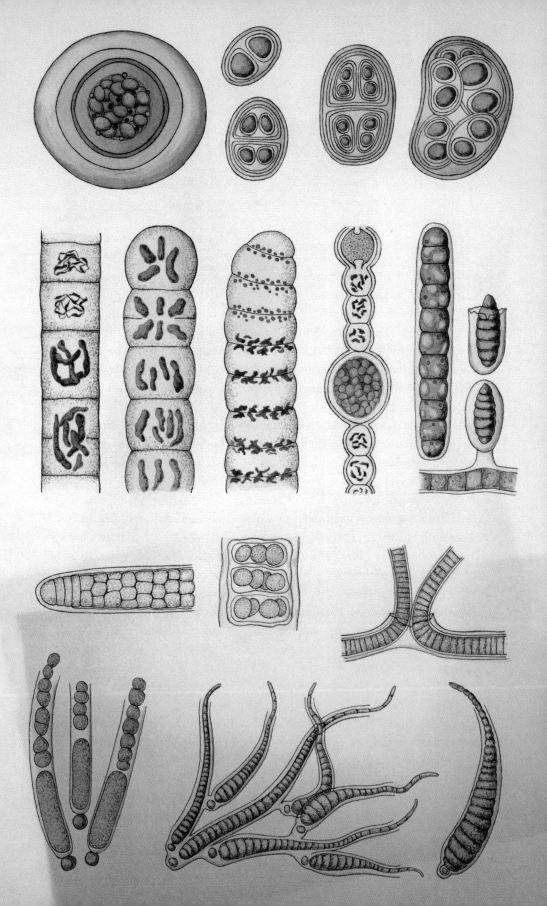

in shallow, well-lit water, and when coral reefs lose their access to light, they die, because their algal partners can no longer photosynthesise. Algae can make a living in many unexpected places: on tree trunks, in the fur of animals, in hot springs and inside banks of snow.

If algae are not plants, they are certainly the ancestors of plants. Photosynthesis changed the world: and eventually allowed small organisms to move from the water to the land. The algal group called stonewort is generally – though, again, it depends on who you ask – considered the ancestor of land-plants. That is a process that began, it's generally reckoned, 475 million years ago.

They can, then, be regarded as the beginning of life, the without-which-not of all large-scale life forms on earth, us included. They have continued to exist in their many different and often unrelated forms for many millions of years, and they live and they die, and when they do so, they generally sink. And that has given us oil.

Not all at once, of course.

It begins with plankton: the stuff that mostly floats in the top of the sea. This was once the dominant life form of the earth; it was the stuff from which we are all descended, and it is still, as we shall see in the next chapter, essential to the continuing life of the planet. The photosynthesising plankton absorbs the energy of the sun and stores it as carbon molecules. This in turn feeds tiny creatures that can be considered animals (if you ask the right person). So here is a system of life: part of it getting energy directly from the sun and part of it indirectly, via consumption. And they live and they die and they sink. Now start considering this process over time: try to imagine what unimaginable lengths of time are like; the sort of time it takes the monkeys to produce a rough draft of Hamlet.

The remains of the floating plants and animals become part of the economy of the earth, buried beneath mud, sand and rock. The immense forces of the planet operate on these fallen minutiae: enormous pressure and high temperatures that come from deep beneath the surface. This, along with the lack of oxygen, transforms the once-living particles into a waxy stuff called kerogen. Add to this more heat, more pressure and more time and this produces hydrocarbons of different types, by a process called catagenesis. One of these types of hydrocarbons is crude oil, which moves through porous rock away from areas of high pressure to low, like an infinitely slow model of a weather system. Often this makes the oil flow upwards, in an uncanny and counter-intuitive fashion. It can sometimes reach the surface, but the days of easily accessible deposits are mostly in the past. In 1949 the average depth of an oil well was about 3,500 feet (1,000 metres); by 2000 it was 19,000 feet (6,000 metres) and the deepest 40,000 feet (12,000 metres) – deeper than Everest is tall. Oil drilling often covers huge horizontal distances as well.

Opposite World-drivers: drawing showing the structure of blue-green algae.

This, then, is the stuff that powered humanity through the twentieth century and beyond, allowing the pace of alteration to the planet to move at a pace beyond easy imagining. It has, as we know, come at considerable cost: the burning of fossil fuels releases gases into the air which retain the heat of the earth, rather than allowing it to escape into space. So-called greenhouse gases are a good thing, because without the heat they have retained throughout the planet's history, life would not have been possible, certainly not life as it exists today. But with the release of greater quantities of these gases by the burning of fossil fuels, the pace has changed, the equilibrium has been lost and the earth is heating up. It is a crisis that seems to us humans to be advancing in relentless slow motion, but in geological terms it is a snap of the fingers.

There was a crisis connected with drastic and sudden temperature change at the end of the Permian Era, 250 million years ago. It is normally accepted that this was caused by a massive series of volcanic eruptions, which sealed in the earth's heat and caused a drastic rise in the climate. The earth recovered all right, as we can bear witness: the snag is that it took 20 million years to do so.

We still live in the age of oil, as most of us attest every time we make a journey. The date for when we run out of oil to burn is another question capable of a thousand answers; one alarmingly precise estimate puts it at 2067. Our plans to phase out fossil fuels are full of loud words and little action, but it looks as if it will be forced on us soon enough, or on our children and grandchildren, who may not be in the mood to thank us. We are beginning to make changes. We have atomic power, which is expensive and dangerous. We also have the so-called renewables. These include biofuels, as we saw in the previous chapter. They also include various methods of harnessing the power of wind and water.

The development of renewables is essential for the long-term survival of human society in the form that it exists today, and also of the ecosystem it occupies: i.e. the planet earth. The problem is that the people who run human society – i.e. politicians and big business – invariably prioritise the short-term: the electoral term. The fiscal quarter. The way we resolve this clash of interests will define the future of our species and of the planet we inhabit.

NINETY-SEVEN
PHYTOPLANKTON

This most excellent canopy the air… why, it appears no other thing to me than a foul and pestilent congregation of vapours.

William Shakespeare, *Hamlet*

Phytoplankton is the planty part of plankton; plankton is the minute specks of life that you can find on and near the surface of oceans and freshwater all over the world. Broadly speaking, it comes in two forms. There is zooplankton, the kind of plankton that needs to eat stuff in order to live, and phytoplankton, which makes its own food by photosynthesis. 'Phyto' is from the Greek word for plant; a neophyte literally means 'a new plant'.

Phytoplankton comprises roughly 1 per cent of the earth's plant biomass, but it accounts for about half the earth's photosynthesis, and therefore about half the earth's production of oxygen. You may not know much about phytoplankton, and you may not have heard of the stuff before, but without it, the planet you live on would be greatly less capable of supporting life.

Phytoplankton is the autotrophic component of plankton: that is to say, it makes its own food. Someone's got to do it: by doing so it enables life in the ocean. This is where the food-web begins: it's what makes a food-web possible. Here's an example, one very short strand in the great and complex food-web of oceanic life: zooplankton eat phytoplankton. The small crustaceans called krill eat zooplankton, and do so with such efficiency that they exist in crowds and clouds of millions. And these clouds feed blue whales. It is a famous paradox that the vastest creature that has ever existed on the planet should choose as its food individuals about the size of your little finger: it is a further and equally instructive paradox that the blue whale is dependent for its existence on photosynthesising shards of life, most of which are too small to be seen without a microscope.

Technically speaking, not every component of phytoplankton is a plant: we are back in define-your-terms territory. Among the wide and varied community – there are about 5,000 known species – there are diatoms, cyanobacteria, dinoflagellates, green algae (so there is some crossover here with the previous chapter) and some eye-pleasing armoured things called coccolithophores. What they have in common is their ability to photosynthesise: and so to make their own food and to produce oxygen.

Essential: marine phytoplankton bloom, Finland.

It's possible to see plankton with the naked eye en masse, as a patch of different colour on the surface of the sea. Phytoplankton is made visible via satellite imagery: as vivid streaks and whorls that pattern the ocean: an effect created by their existence in uncountable billions of individuals. Such a slick can cover hundreds of square kilometres. Views of the earth from space have given us a series of revelations about the nature of the planet we live on, beginning with the first picture of the brave blue marble that was taken by *Apollo 17* in 1972: the picture that gave humanity our first visceral understanding of the finite nature of what we call the world. Oceanic images showing the living patterns of plankton add depth to this initial understanding. They tell us about incomprehensible numbers and levels of complexity beyond the intuitive grasp of the human mind: they also show us the patent fragility of life.

Phytoplankton is about light. That means it can only exist in the well-lit parts of the ocean: on the surface and in the areas just below: technically the euphotic zones, not just where life penetrates, but where there is plenty available. And like plants in the ground, phytoplankton requires nutrients which can be found in most waters: nitrate, phosphate, silicate, calcium and trace amounts of iron. Phytoplankton is prone to seasonal fluctuations: in the northern latitudes there can be a great explosion of growth in the spring; in the tropics a considerable drop-off in the summer months.

The presence of phytoplankton is not invariably benign. In favourable conditions some species can multiply fast; some of these species produce biotoxins for defence. This can create a hostile algal bloom, nicknamed the red tide. It can kill fish, and it can harm and sometimes kill those that eat the fish, including humans. This red tide can also sink to the bottom and decompose, depleting the water of oxygen and creating a dead zone.

That said, phytoplankton is essential to the continuation of life on the planet. It plays an important role in the sequestering of carbon from that atmosphere, which makes it important in the slowing-down of rising temperatures. It's been calculated that every year, phytoplankton takes 10 gigatonnes of carbon from the atmosphere and transfers it to the ocean depths: that's 10,000,000,000 tonnes.

We now know that these pleasing streaks of oceanic colour, these invisible specks of life that hardly anyone has ever seen as individuals, are important to us. Phytoplankton is essential if we want to continue taking food from the seas; it is essential if we want to keep climate change to survivable levels; it is essential if we want to keep breathing. There is a big question then: is the phytoplankton of the world in good health?

Some studies claim to show that phytoplankton density has reduced over 100 years, but that's open to question; we don't have phytoplankton studies that go back that far, so it's hard to compare. But there are concerns to do with the issues that affect phytoplankton: salinity, water temperature, depth, acidification, wind and the numbers of predators. Satellite imagery from the past decade appears to indicate a small decrease.

One of the problems that come with warmer oceans is that higher temperatures interfere with the process of vertical mixing: that is to say, the extent to which water of a shallower depth can mix with water lower down. You can experience this yourself with a clean dive into warm water: you will feel the warm top layer instantly followed by much colder stuff underneath as you plummet down. There is no intermediate zone because the two layers find it hard to mix. One side effect of the lack of vertical mixing is that the nutrients needed by the phytoplankton community become less readily available.

In phytoplankton we have something we know very little about, apart from the fact that it's essential. Its beauties are elusive to most of us, and for that matter, so is its very existence. Our lives depend on stuff that's hard to understand and impossible to see with the naked eye.

NINETY-EIGHT
BAOBAB

Wisdom is like a baobab tree; no one individual can embrace it.

African proverb

It would have looked strange anywhere. But where it stood, on a slight rise in the middle of the Kalahari – a place where it seemed that nothing else was growing for hundreds of miles – it looked impossible. Its girth was a thing of wonder in itself: it would have taken a dozen of us, hand in hand, to encircle it. Its unnaturally thick branches gave it a look of savage power, a survivor from a forgotten tree-world.

This was Chapman's baobab, a national monument in Botswana, named for James Chapman, the explorer. He came this way: how could he not? The tree is the only landmark in many million acres of land. In its shade pioneers stopped and rested. They left letters for each other in its hollows, sometimes the last that was ever heard of them. They carved their initials in its bulk and they are still visible. They chose the tree because, even when two centuries younger, it was the most imposing thing you would pass in many weeks of travelling.

I sat beneath it with a small party and a strange spell came over us. We talked with an almost unnatural freedom, discussing things unmentionable in other circumstances. We shared near-death experiences and drinks. The strange compulsion of the desert had us in thrall: and was released by our intimacy with this quite extraordinary tree.

The hottest day in Botswana's history came on 7 July 2016, perhaps a decade after my visit. It killed Chapman's baobab. In the past decade, nine of the thirteen oldest baobabs in Africa have died.

There are eight species of baobab, all in the genus *Adansonia*. Six are endemic to Madagascar, one to Australia and one to the African mainland below the Sahara. The more widespread of these, *Adansonia digitata*, is the main subject of this chapter. These baobabs are marked by a massive barrel trunk and a rather compact crown. The wood is spongy and fibrous, enabling it to store around 26,000 gallons (32,000 US gallons, 120,000 litres) of water. They are, then, well adapted for drought, but they are not exclusive to deserts, found across the savannah, though never in great numbers.

They can live to a great age. We have a tendency to exaggerate such matters, as if the plain facts were not enough for wonder. The Panke baobab in Zimbabwe,

Trees with a story to tell: Lions by the Baobab, c.2013, by Joseph Thiongo (b. 1967).

which died in 2011, was calculated at 2,450 years old, and was then the world's oldest living angiosperm. (Some species of conifers, which are gymnosperms, live longer, as we have seen.) Chapman's tree was probably around 1,000 years old at its death, but it's hard to age them; the fibrous nature of their interior means that there are no annual growth rings to count.

Their extraordinary shape compels the attention: in many places baobabs are known as upside-down trees, because their branches look like roots. They produce very large aromatic flowers which bloom for about fifteen hours, mostly at night, and are pollinated by bats, bushbabies and night-flying moths. Young baobabs look spindly, very different in shape from the vast mature individuals, which can have trunks 30 feet (9 metres) in diameter. This means that the youngsters are easily overlooked, giving the idea that the only baobabs in existence are incalculably ancient. That enticing notion is one of the reasons the trees have such immense cultural importance, for they are considered to be the hosts of ancestral spirits.

The tree is also a reminder of God's anger. There are stories that tell of the time that God hurled the lush and verdant tree into the sky. It landed upside down with its roots on show, its glories forever buried. In some stories this was a punishment for the tree's pride in its own beauty. In another interpretation, God was angry because humans were wrecking his creation. He instructed the baobab to bloom only at night, when no human could see. In Siavonga, a small market town in Zambia, there lived a woman who had traded recklessly in the earth's treasures. She was killed by a tree that was thrown by the angry God; it landed on its side and crushed her. The tree, a national monument, is called Ingombe Ilede, which means sleeping cow, for the bizarre shape.

A baobab is not only imposing, it is capable of living on after suffering immense damage. It can survive being ring-barked, which is lethal to most trees because the outer layers of the trunk transport water to the rest of the tree; a baobab can perform this feat right through to the middle. The tree can carry on even after great gouges have been made in the trunk. This facility for self-repair gives the tree yet more meaning: there is a belief that if the umbilical cord of a newborn child is wrapped and dried in the mother's old sanitary cloth and then pushed into a hole in a baobab, the child will grow fat and strong, resistant to diseases, and will live a long, full life.

Baobabs often grow hollow, creating commodious shelter to many species. Once, while seeking a monitor lizard, I disturbed different tenants, a hive of bees; I was lucky to escape with only two stings. Such trees, such shelters, are known and used by human hunters who can't make it back to the village for the night. An ancient baobab will often be the centre-point of a village: and it's a place where important meetings of the elders are held. These go all the better for the presence and the wisdom of the ancestors in the baobab: the trees are emblems of a rich, continuing culture. This sense of the power and meaning of the baobab transcends tribal beliefs, as the response from outsiders has often shown. The American writer Peter Matthiessen wrote a book that was basically a love letter to sub-Saharan Africa, and he named it for the baobab: *The Tree Where Man Was Born*. Everyone who sits beneath a mature baobab must consider immensities of time.

The baobab is said to have 300 different uses. Its fibres can be plaited into a string or made into baskets. Pound for pound, the fruit contains six times more vitamin C than an orange. The leaves are rich in iron. The seeds can be used as coffee substitute; they can also be sucked and the vitamin-rich powder that covers them can be relished. The seeds are used for a traditional board game called nsolo. They yield a good oil for cooking and cosmetics. You can make beer from the fruit. The tap root of a baobab seedling can be eaten like a carrot. You can get a red dye from the roots.

The baobab has many medicinal uses. Its leaves, bark and seeds can be used to treat diarrhoea, toothache and malaria. When my friend the botanist Manny Mvula was a boy in Zambia, a skin rash went round his school from shared towels. With the advice of a friend's uncle he treated this with a paste of baobab root and Vaseline and it cleared up very quickly. More remarkably, his step-brother's children were diagnosed with sickle-cell anaemia in hospital. Back at the village they were given a mixture of baobab root and leaves with red mahogany bark; when they made their next visit to hospital, to considerable surprise it was found that regeneration of fully formed red blood cells was taking place.

In the course of his botanical survey of the Luangwa Valley with Bill Astle, Mvula once found a baobab with an elephant tusk embedded in it; it had broken off while the elephant was gouging the tree in search of water and nutrition. Exploitation by elephants has been part of life for baobabs across the millennia, but of late the pace has hotted up. There are two reasons for this. The first is that both elephants and an ever-expanding human population need a lot of space, and the humans usually win: elephants and their migration routes get squeezed and elephants are confined to smaller spaces. The second is that with climate change there are longer droughts, changing weather patterns, and there is less water around. The combination of these factors means that elephants are turning to their emergency water supply more often: baobabs. As a result, baobabs in places like Mana Pools in Zimbabwe are dying at an unprecedented rate. Even without elephant intervention baobabs are dying, and climate change is at the heart of this. Baobabs, once the perfect emblem of the ancientness and longevity of the human population who revered them, are now a vivid demonstration of the impermanence and fragility of life on this planet.

NINETY-NINE
ROSY PERIWINKLE

*The art of medicine consists in amusing the patient
while nature cures the disease.*

Voltaire

In chapter after chapter of this book we have come across plants prized for their medicinal properties. Some of these claims seem reckless and extravagant: plants that are supposed to cure everything, true panaceas. Other claims are relatively modest: plants that relieve the symptoms of some humdrum disorder. There are plants said to offer immortality and plants that make a cold easier to bear. In the previous chapter there were suggestions that baobab can cure – or at least help in the treatment of – both a mild skin rash and potentially lethal sickle-cell anaemia. We are entitled to wonder how much of this is superstition, how much the placebo effect and how much genuine disease-busting biochemistry. There is often polarisation involved in our response to such questions: uncritical belief on one hand and over-cynical rationalising on the other.

Many of the celebrated individuals in these pages combined the profession of botanist and pharmacist. It is an ancient connection: the idea that they can be separated is comparatively recent, and is mostly a phenomenon of Western culture. Manny Mvula will use technical botanical terms and explain the traditional uses of the plants encountered in the course of a walk through the bush, implying that the two disciplines have no need to go through a hostile divorce. Plants can be effective at many things: they can kill you and they can get you high, so there is no reason they can't cure you, as everyone one who has taken an aspirin (see Chapter 5) knows.

So let us turn to a plant endemic to Madagascar called the rosy periwinkle, also Madagascar and cape periwinkle, bright eyes, graveyard plant and old maid. It is an evergreen herbaceous plant that bears a pretty five-petalled flower. All parts of it are highly toxic. It is cultivated as an ornamental plant, prized because it does well in dry and poor conditions. It can't survive below 5°C (41°F), so in temperate places it's used as a bedding plant: that is to say, cultivated under glass and planted out in the warmer months.

Opposite *Cure for cancer: rosy periwinkle, 1900, by Kōno Bairei (1844-95).*

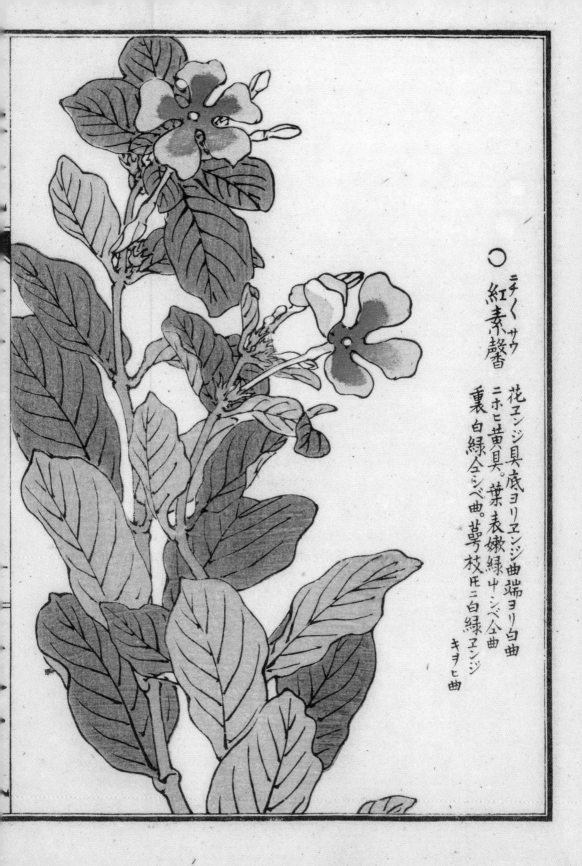

○
紅素馨
ニチくサウ

花ヲンジ具底ヨリヱンジ曲端ヨリ白曲
ニホヒ黃具。葉表嫩綠中ニシベ仝曲
裏白綠仝シベ曲。萼枝𠮷ニ白綠ヱンジ
キヲヒ曲

Inevitably, or perhaps I mean naturally, it has its traditional uses, and has been used to treat diabetes and malaria. It spread out from Madagascar early, and was used in Mesopotamia around 2600 BC, and later in India and China. It's possible that it would have remained a minor curiosity for folklorists, so far as the Western world is concerned, but in the 1950s the plant was examined closely in laboratory conditions to measure its effectiveness against diabetes. Nothing earth-shaking was discovered in this department. What they did find was that the plant produces substances that can be dramatically effective in the treatment of certain kinds of cancer. One of the alkaloids, vinblastine, is a treatment for Hodgkin's disease, which attacks the immune system and is a killer, particularly with young adults. The second, vincristine, is used to treat childhood acute lymphocytic leukaemia. Before this was available, diagnosis was more or less a death sentence: about 10 per cent survived. Now it is about 90 per cent.

In its natural habitat in Madagascar, the rosy periwinkle is an endangered species. But it's widely cultivated in Australia, Malaysia, India, Pakistan and Bangladesh. In parts of Australia it has become an invasive plant and is categorised as a noxious weed. The value of the annual trade in this plant has been calculated at £750 million. None of this goes to Madagascar. The pharmaceutical companies that have patented substances that come from this plant have been accused of biopiracy.

You need a lot of them. It takes 500 kilos (1,100 lbs) of plant matter to create a single gram (.04 ounces) of the desired alkaloids. In 2018 at the John Innes Centre in Norwich the genome of the rosy periwinkle was sequenced, and it's possible that this will lead to a cheap and effective method of synthesising the drug. But right now we still need the living plants.

In Madagascar the clearance of natural spaces for farming continues. In this country of desperate poverty, slash-and-burn agriculture is prevalent and an ever-growing threat to the country's future. It's worth asking what would have happened if the rosy periwinkle had gone extinct before it reached the laboratories… and we can extend the question to ask how many plants that might have been immensely useful to humanity have already gone for good.

About 5,000 plants have been tested for alkaloids that would be useful in medicine. That's quite a lot – but there are still more than 200,000 to go: plants that might have properties that would save lives and perform all manner of marvellous things. They have not been tested in the laboratory, but every single one of them has gone through a far more rigorous testing process than any scientist can devise: that of natural selection. Plants develop their defences and their strategies on the way to becoming ancestors: in this way they are tested across uncountable millennia of competition. Those that survive can only be genuine achievers. The evolutionary biologist Edward O. Wilson wrote: 'They are better than all the world's chemists at synthesising organic molecules of practical use.'

When you think about it, you see that this has to be the case. And what plants have developed to protect themselves can be used by humans for our own purposes. Among the drugs we have looked at in this book are aspirin, caffeine, cocaine, codeine, digitoxin, morphine, penicillin, quinine and, in this chapter, vinblastine and vincristine. In the United States a quarter of all drugs prescribed by pharmacists come from plants.

There are many, many plants with properties still untested by science. There are many, many arguments for maintaining the biodiversity and the bioabundance of the planet we live on. Here is one that doesn't require imagination or altruism or breadth of thinking or understanding of deep time. There are plants out there that will keep death away, at least for a while. As we destroy the planet's natural resources, so we destroy possibilities for the future.

DIPTEROCARP

Under the greenwood tree
Who loves to lie with me…

William Shakespeare, *As You Like It*

Ever since we invented agriculture 12,000 years ago we humans have made a sharp distinction between what is tame, controlled and part of the human landscape, and what is wild, untamed and beyond our control. But in the course of history our attitude to wild places has changed. Once they were mighty,

threatening, forbidding, places only for the most courageous; it was universally accepted that nature was coming to get us, forever waiting to destroy all we have achieved as a species, everything that we dared to call civilisation. Now we increasingly accept that wild places are fragile, beautiful, on the verge of being lost forever. Once wild places were so useless to us that we called them wastes: now we are beginning to accept that they are of crucial importance to humanity. One landscape in particular sums up this shift: rainforest.

So now, as we come to the end of this book, we can mark our century by turning to the classic rainforest tree. It stands taller even than the tall trees of the surrounding forest, an emergent that raises its head above the canopy and dominates the landscape. You can tell it from the others even from the ground, in the dark avenues where only 2 per cent of the sunlight can reach. The tree is

Photograph of mist and low cloud hanging over lowland dipterocarp rainforest, just after sunrise in the heart of Danum Valley, Sabah, Borneo. Photographer: Nick Garbutt.

different because it has buttresses: roots that start above the ground, sometimes higher than your head, stretching out with immense certainty and conviction to give the giant the added support that its greatness demands.

This is the dipterocarp: a group that contains 16 genera and nearly 700 species. They are not all buttressed emergents, but these are the ones that stick in the human mind. They *are* rainforest: as mighty as they are fragile.

The name means winged fruit: the seeds are distributed by aerodynamic structures that allow them to exploit the wind, so that at least some of them will travel a fair distance from the parent plant. They are found all over the world about the middle parts, a pantropical group, though they are seen at their best in Borneo. In these Bornean forests 22 per cent of all trees are dipterocarps, and there are 269 species on the island. Of these, 162 are found nowhere else, and over 60 per cent are threatened with extinction. The record for height is 305 feet (93 metres) attained by an individual of the species *Shorea faguetiana*.

They are colossal, and that makes them a colossal resource to many other species of the forest, a central part of the greatest living community on earth. When a huge tree like this comes into fruit, it becomes instantly attractive to all the many creatures that can feed there. Some trees need to have their fruit consumed, so that the seeds can be distributed in the droppings of the consumers, which is why their fruit are both decorative and tasty. Dipterocarps work on a different principle: the wind does the job of distribution, and so excessive consumption of their unostentatious but nutritious fruit would be a serious problem.

They combat this by holding off production for some years and then coming up with a bumper crop. This works because the fruiting is synchronised with other trees of the same species. They all produce their fruit at the same time, what is technically known as a mast year. (Oak trees operate a similar strategy on a less gorgeous scale.) This creates a superabundance of fruit: more than the jungle community can consume. It is a strategy of predator satiation. It has a payback: the tree might die – struck by lightning, perhaps, a common end of emergent trees in the rainforest – before it can take part in the next mass fruiting.

The wood of the dipterocarp species has long been used by humans. The biggest trees are naturally the most desirable targets for loggers of the legal and the illegal kind. Dipterocarps are slow-growing and produce a dense timber, resistant to insects and fungi and capable of lasting many years. Dipterocarps have been selectively logged and also been part of many events of mass felling, to create pasture, to make room for crops, for human development.

Rainforest covers at best 6 per cent of the land surface of the planet but contains more than 50 per cent of all species in all taxa: some say much more than that. Destruction of this is not merely a shame: it is a catastrophe for humanity. Biodiversity is not just quite interesting: it is the way that life works and continues. Diversity is resilience and vigour. Life operates as a complex network of dependencies, and its strength lies in its complexity. Life on earth has often been compared to a web: break a single strand and the whole structure is weakened. There is also the rivet-popper hypothesis: an aeroplane is flying through the air and it loses a rivet. So what? It has lots of rivets. It loses another, and another, and another... and if it carries on losing rivets it will eventually fall out of the sky. One rivet: one species. Rainforest has more species than anywhere else: so we are getting rid of rivets at a frightening rate.

Biodiversity has become a much-used word in the course of the twenty-first century. It was coined by the scientist and writer Raymond F. Dasmann in 1968, and it was used in *The Diversity of Life*, an important and influential book by Edward O. Wilson published in 1992. Since then it has entered the currency, just as the word rainforest did at round about the same time.

Our understanding of the phenomenon is much older than Genesis, but we can turn to Genesis for conformation: 'And God said, Let the earth bring forth grass, the herb yielding seed, and the fruit tree yielding fruit after his kind, whose seed is in itself, upon the earth: and it was so.' The account continues with the creation of the animal kingdom and then humanity. It all begins with the creation of the plants, and all life on earth still depends on plants. We have always understood life as a collection of many, many different species.

If we have always had such a concept, why did we only need to find a word for it at the end of the twentieth century? It was because we had always taken it for granted: we knew, without needing to be told, that the earth and the seas were teeming. It was the way life is. It was only when we began to realise the extent of our destructiveness that we needed the word.

The dipterocarp, quite literally above all other trees, is the symbol, not just of rainforest, but of all biodiversity and all human folly. I have stood between the buttresses of living emergent dipterocarps and felt gloriously humbled; I have walked across places where dipterocarps once stood and felt helpless sadness. And I have visited a project in Borneo called Hutan where forest reclamation continues to this day. I was invited to plant a tree, and, most unhandily, I did so. The hands that wrote this book also planted a dipterocarp. I know which was the greater contribution to the future of humanity and of all life on this planet.

EPILOGUE

The day began in a rose garden, but it was a garden no human hand had planted or pruned or choreographed. These wild roses were growing either side of a shallow stream that flowed through flat land at the foot of a mountain in the Caucasus: pink roses and white roses, backlit by a low sun: a place so close to Eden that I half-expected to meet the naked couple themselves, perhaps sharing the glorious fruit of this region.

I was there with an Armenian NGO called Foundation for the Preservation of Wildlife and Cultural Assets. I was working with the British-based World Land Trust on a plan to conserve and expand the Caucasus Wildlife Refuge. It seemed rather early in the day to achieve perfection, and it turned out that we had only just started.

We climbed with the sun: higher and higher up the mountains in a bucking vehicle. Soon we were travelling through alpine meadows, places no one tended but the sparse and scattered domestic stock – small cows, sheep and tough, bony little horses – along with bezoar goats and mouflon or wild sheep.

It was a place of reckless colour, a landscape too rich for easy understanding, a million million flowers competing against each other in beauty, each one seeking to attract the pollinators at the peak time of the year for warmth and sun and life. It was now or never: and all around me the mountainside was opting for now. Often the colours broke free of the flowers and filled the air, for the place was dizzy with butterflies and other gorgeous pollinating insects. Once or twice a sky-filling lammergeier flew over.

On we climbed, on: could we find – sordid thought, inspiring thought – the money to secure this place and look after it for years to come? At last we were at the summit, cold air and all the lushness gone, the occasional rattle of stones knocked aside by the bezoars on their precipitous rocks. I sat on the edge, feet resting on a convenient ledge; between my boots I saw another lammergeier.

And all round me were the plants of limestone pavements: a thrilling range of colours, and yet in these hard conditions none of them more than a couple of inches tall, fecundity in miniature, showing me yet another way in which plants can make a living, another way in which life can make more life.

The long hard-climbing day was a revelation of plenty, of beauty, of biodiversity and bioabundance. Here was the glory of the world. Look at these plants. Look at this kingdom of plants. Look at this planet and its uncountable plants. We owe them everything.

INDEX

PICTURE CREDITS

© Look and Learn; **Pg 177** © Barbara Philip, www.africanpainting.com; **Pg 180-181** Alamy / agefotostock; **Pg 185** akg-images / André Held; **Pg 188** Histoire Generale des Antilles Habitees par les Francois Vol 2 by Jean Baptiste Du Tertre (1610-1687), published 1667. Bridgeman Images; **Pg 191** akg-images / Mark De Fraeye; **Pg 195** Bridgeman Images / © Look and Learn; **Pg 199** Livre de maison Cerruti, *Ail* / Manuscrit, fin du 14e siècle, Vienna Austrian National Library. Cod. ser. nov. 2644. Alamy / Album; **Pg 200-201** Alamy / Peter Horree; **Pg 204** Getty Images / Bettmann; **Pg 208** Alamy / History & Art Collection; **Pg 212** Alamy / Heritage Image Partnership Ltd; **Pg 216** Glenn Marshall, *Easy Grow Japanese Knotweed seeds*, 2015. Cartoonstock / © Glenn Marshall; **Pg 217** Science Photo Library / Lizzie Harper; **Pg 220** Bridgeman Images; **Pg 224** © The Estate of Frank Moss Bennett. All rights reserved. DACS 2022. akg-images / © Sotheby's; **Pg 226-227** Marianne North (1830-1890), *Tea Gathering in Mr Holles Plantation at Garoet, Java*. Copyright © RBG KEW; **Pg 229** Bridgeman Images / © Harris Museum and Art Gallery; **Pg 233** Bridgeman Images / © Historic England; **Pg 237** Getty Images / Universal Images Group; **Pg 241** Getty Images / Fine Art; **Pg 246** Bridgeman Images / © Bristol Museums, Galleries & Archives / Given by Miss Margery Fry, 1935; **Pg 248** Bridgeman Images / © Mondadori Electa; **Pg 252** Alamy / Heritage Image Partnership Ltd; **Pg 256** Alamy / Chronicle; **Pg 260-261** Alamy / The Print Collector; **Pg 265** Alamy / CPA Media Pte Ltd; **Pg 269** Alamy / Hercules Milas; **Pg 270-271** The Metropolitan Museum of Art, New York, Robert Lehman Collection, 1975. Acc No: 1975.1.74; **Pg 275** Bridgeman Images; **Pg 279** akg-images / De Agostini Picture Lib. / A. Dagli Orti; **Pg 282-283** Getty Images / De Agostini Picture Library; **Pg 288-289** akg-images; **Pg 291** Bridgeman Images / © Look and Learn; **Pg 294** Alamy / Stocktrek Images, Inc.; **Pg 297** Bridgeman Images / © Archives Charmet; **Pg 301** Wikipedia Commons / https://commons.wikimedia.org/wiki/ File:Hieronymus_Bosch_-_The_Garden_of_Earthly_Delights_-_Prado_in_Google_ Earth-x2-y0.jpg; **Pg 305** Alamy / Granger Historical Picture Archive; **Pg 309** Alamy / Pictures Now; **Pg 313** Alamy / Album; **Pg 317** Jean-Léon Gérôme (1824-1904), The Tulip Folly,1882. Walters Art Museum. Gift of Mrs. Cyril W. Keene, 1983. Acc. No: 37.2612 **Pg 321** British Museum / © The Trustees of the British Museum; **Pg 325** Getty Images / DEA A. Dagli Orti, **Pg 329** Bridgeman Images; **Pg 333** Alamy / Florilegius; **Pg 337** akg-images / Erich Lessing; **Pg 339** Getty Images / Photo Josse/Leemage; **Pg 342** Alamy / Les Archives Digitales; **Pg 345** Alamy / Pictures Now; **Pg 349** Getty Images / DEA Picture Library; **Pg 352** akg-images; **Pg 357** Bridgeman Images / Photo © Christie's Images; **Pg 361** Getty Images / DEA / M. Seemuller; **Pg 364** Alamy / Album; **Pg 368** Bridgeman Images / Bourne Gallery, Reigate, Surrey; **Pg 372** Science Photo Library / Dr Keith Wheeler; **Pg 375** Bridgeman Images / Photo © Whitford & Hughes, London, UK; **Pg 378-379** Alamy / Contraband Collection; **Pg 383** Bridgeman Images; **Pg 387** Dreamstime / Eve Mazur; **Pg 390** Bridgeman Images / © NPL - DeA Picture Library; **Pg 394** Science Photo Library / NASA; **Pg 397** True African Art / Joseph Thiongo; **Pg 401** Alamy / Florilegius; **Pg 406-407** Nature Picture Library / Nick Garbutt; **Pg 408** Nature Picture Library / Paul Harcourt Davies; **Pg 431** David Bebber.

Every effort has been made to find and credit the copyright holders of images in this book. We will be pleased to rectify any errors or omissions in future editions.

ACKNOWLEDGEMENTS

No matter how much the author wants to hush it up, a book like this is always a team effort: so thanks to everyone who has made *100 Plants* possible. First, a special *zikomo* to Manny Mvula, botanist and wildlifer, for his invaluable assistance on the African species. Thanks to Ralph Mackridge and Jill Millar for much help and important suggestions. At Simon & Schuster, thanks to Ian Marshall for setting up the project and to Laura Nickoll for making it happen. Thanks to Victoria Godden for the editing, to Keith Williams, the designer, and to Mary Chamberlain for proofreading; also to Liz Moore, who did the picture research and never gave up. I'm deeply grateful to Sara Oldfield for casting an expert's eye over the manuscript. At Georgina Capel Associates, thanks as always to George, and also to Irene Baldoni and Rachel Conway. And of course, I couldn't have done it at all without Cindy, Joseph and Eddie back home in Norfolk.

Good start: the author planting a dipterocarp seedling at the Hutan project, Malaysian Borneo in 2012. Photographer: David Bebber.

First published in Great Britain by Simon & Schuster UK Ltd, 2022

Editorial Director: Ian Marshall
Design: Keith Williams, sprout.uk.com
Project Editor: Laura Nickoll
Picture Researcher: Liz Moore
Proofreader: Mary Chamberlain

1 3 5 7 9 10 8 6 4 2

Simon & Schuster UK Ltd
1st Floor
222 Gray's Inn Road
London WC1X 8HB

www.simonandschuster.co.uk

Simon & Schuster Australia, Sydney

www.simonandschuster.com.au

Simon & Schuster India, New Delhi

www.simonandschuster.co.in

A CIP catalogue record for this book is available from the British Library

Hardback ISBN: 978-1-3985-0548-3

Ebook ISBN: 978-1-3985-0549-0

Printed in Bosnia and Herzegovina

MIX
Paper | Supporting
responsible forestry
FSC
www.fsc.org   FSC® C118234

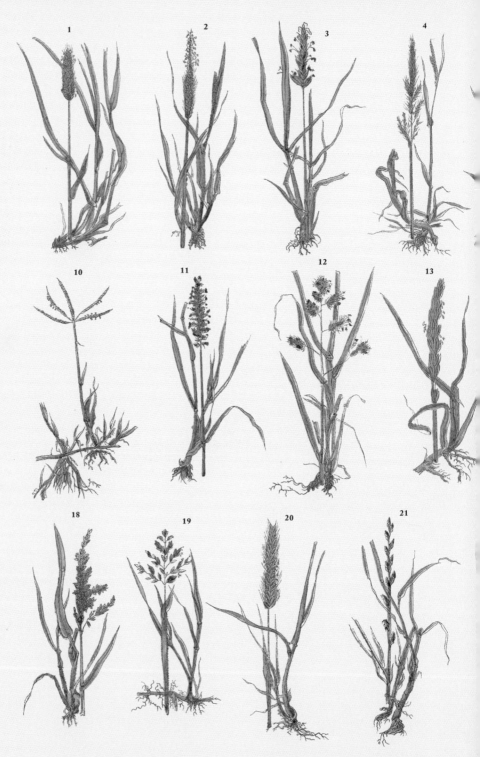

1., ALOPECURUS PRATENSIS, *Meadow Foxtail* ; – 2., AL. AGRESTIS, *Cretan Meadow Foxtail*; – 3., ANTH
6., AVENA PRATENSIS, *Meadow Oat-grass;* – 7., BRIZA, *Bitter Grass* ; – 8., BROMUS MOLLIS, *Bull Gr*
12., DACTYLIS, *Cock's-foot Grass;* – 13., ELYMUS, *Wild Rye Grass;* – 14., FESTUCA OVINA, *Sheep's Fescue*
17., FEST. LOLIACEA, *Tall Fescue Grass;* – 18., HOLCUS LANATUS, *Common Velvet Grass;* – 19., HOL
22., PHLEUM, *Field Grass;* – 23., POA ANNUA, *Annual Meadow Grass;* – 24., POA PRATE